GREENING DEVELOPMENT IN THE PEOPLE'S REPUBLIC OF CHINA

A DYNAMIC PARTNERSHIP WITH THE ASIAN DEVELOPMENT BANK

OCTOBER 2021

ASIAN DEVELOPMENT BANK

ADB

Notes:
In this publication, "$" refers to United States dollars. ADB recognizes "China" as the People's Republic of China.

On the cover: The PRC invests in bus rapid transport systems, wind power, biodiversity conservation, and sustainable water resources management to promote green development (photos by ADB, Arthur Hanson, and the Zhaotong Yudong Reservoir Management Bureau).

Cover design by Edith Creus.

CONTENTS

TABLES, FIGURES, AND BOXES

FOREWORD

Throughout 3 decades of our collaborative journey, the Asian Development Bank (ADB) and the People's Republic of China (PRC) have sought to balance economic progress with environmental sustainability. The country's efforts to achieve carbon neutrality by 2060 and the renewed focus on green development in the 14th Five-Year Plan bring even greater focus on the need for collective action for nature-positive recovery.

This report shares the narrative of the PRC's efforts to promote green development and ecological restoration as well as accomplishments resulting from its partnership with ADB.

A nature-positive recovery can provide opportunities for economic growth, societal well-being, biodiversity protection, and climate stabilization. This publication traces how, over the years, the PRC has adopted valuable experimentation and piloting to create new tools for advancing green growth, eco-efficiency, protection of nature, pollution control, and use of information technology and digitalization. These tools are now being employed for achieving synergies across sectors and for building new economic opportunities for businesses contributing to the green economy. The PRC's policies to promote ecological restoration, rural vitalization, and green finance can address both environmental protection concerns and regional disparities.

The PRC's cutting-edge initiatives on environment and development can provide important lessons and replicable practices not just for ADB and its developing member countries, but also for the global community. In line with the ADB–PRC partnership on rural vitalization, ADB is introducing climate-smart farming practices in the Yangtze River Basin to increase production, avoid environmental degradation, and enhance the skills of farmers to adopt sustainable agricultural technologies. ADB is also transforming cities to become greener and more livable through integrated, low-carbon solutions. These include constructing sponge city green infrastructure and bus rapid transit corridors that are integrated with green spaces for walking and cycling, adopting climate-resilient and smart urban water management systems as an effective approach to prevent flooding, and improving capacities of local governments in healthy city planning.

The PRC has leveraged its partnership with ADB to advance progress on green development investments, policy dialogue, technical assistance, and knowledge solutions. ADB's own knowledge and capacity have increased in preparing, financing, and implementing green development projects in the PRC across multiple sectors, including renewable energy, ecological restoration, water resources management, biodiversity protection, and urban development.

As we move forward, ADB and the PRC will manage and harness this cooperation and collaboration to help transform a future with economic, social, and environmental benefits that pursue high-quality green development.

Ahmed M. Saeed
Vice-President
East Asia, Southeast Asia, and the Pacific
Asian Development Bank

FOREWORD

This report traces the partnership between the Asian Development Bank (ADB) and the People's Republic of China (PRC) since the early 2000s. The partnership's focus has shifted from traditional projects to comprehensive programs that promote environmental protection, sustainable rural and urban development, and climate change mitigation and adaptation. This evolution presented challenges due to the PRC's fast economic growth, which require more sophisticated solutions.

This dynamic partnership has been driven by combining ADB's project financing with technical assistance. ADB's policy advice and knowledge support for more than 2 decades were critical in developing the PRC's pioneering concept of ecological compensation—a tool to protect the environment and livelihoods under competing demands for natural resources. ADB's support has helped formulate a national regulation on eco-compensation, to be circulated in 2022.

ADB has assisted the PRC in designing projects and programs that pursue a holistic approach to transform the rural economy. For example, ADB's support for the national rural vitalization strategy promotes comprehensive interventions to reduce social disparities among regions and within provinces.

A major innovation was the design of multisector programs with integrated solutions. To support the Yangtze River Economic Belt (YREB) development plan, ADB packaged lending and nonlending assistance with a crosscutting framework within the entire basin. With investments and technical assistance since 2015, the focus in the YREB is on institutional strengthening and policy reforms, natural resources management, ecosystem restoration and pollution control, low-carbon transformation, and sustainable agriculture. This ecological corridor concept will also be applied to protect and restore the Yellow River basin's ecosystems and promote more equitable and sustainable rural–urban development. ADB also designed a multisector program in the greater Beijing-Tianjin–Hebei area to mitigate climate change through improved air quality and reduced carbon dioxide emissions.

The ADB-PRC partnership has scaled up integrated solutions that can tackle increasingly complex development issues. It has provided an opportunity for ADB to learn and showcase the results of more advanced knowledge and policy-based interventions pioneered in the PRC through mutual learning and exchange of best practices. Knowledge and institutional strengthening are central to ADB's operations under the new country partnership strategy for the PRC for 2021–2025. ADB is happy to cooperate with the PRC in sharing valuable lessons and insights from this joint venture.

M. Teresa Kho
Director General
East Asia Department
Asian Development Bank

ACKNOWLEDGMENTS

This report was conceptualized and prepared by the East Asia Department (EARD) under the guidance of James Lynch, former director general, and M. Teresa Kho, current director general. EARD Regional Management Team Xiaoqin Fan, Yolanda Fernandez Lommen, Sujata Gupta, Thomas Panella, and Sangay Penjor; and Qingfeng Zhang, chief of rural development and food security (Agriculture), Sustainable Development and Climate Change Department, provided strategic direction.

A core team led by Yumiko Tamura and Silvia Cardascia managed the overall production of the report. Core team members Marga Domingo-Morales and Charina Cabrido provided technical inputs, coordination, and research support. Significant technical contributions were made by ADB sector and thematic experts from EARD as well as other regional and knowledge departments. They are Bo An, Mark Bezuijen, Xueliang Cai, Silvia Cardascia, Jinqiang Chen, Nicolas Dei Castelli, Warren Evans, Mingyuan Fan, Gloria Gerilla-Teknomo, Annabelle Giorgetti, Robert Guild, Dongmei Guo, Jan Hinrichs, Anqian Huang, Hisaka Kimura, Shingo Kimura, Yoshiaki Kobayashi, Kang Hang Leung, Susan Lim, Marzia Mongiorgi-Lorenzo, Xuedu Lu, Zhiming Niu, Rabindra Osti, Frank Radstake, Arun Ramamurthy, Stefan Rau, Nogendra Sapkota, Hsiao Chink Tang, Takeshi Ueda, Au Shion Yee, Lei Zhang, Yun Zhou, and Yaozhou Zhou. Their write-ups and inputs were significant in structuring the different chapters on environmental protection and ecological conservation, rural economy, green livable cities, and climate change mitigation and adaptation. Arthur Hanson contributed inputs and write-ups during the drafting stage, and also peer reviewed the manuscript.

We are grateful to the following who have peer reviewed the report: former ADB officials Javed Mir, Robert Wihtol, and Xianbin Yao; advisor of the China Council for International Cooperation on Environment (CCICED), Scott Vaughan; and Chief Executive Officer of Climate Bonds Initiative, Sean Kidney.

We also thank the following for their contributions: Maria Theresa Mercado edited the manuscript. Akiko Terada-Hagiwara, Hsiao Chink Tang, Sophia Castillo-Plaza, and Xuan Rong from EARD, and ADB's Department of Communications helped in the publication of the report.

Case studies and other information presented in this publication were derived mostly from ADB's sovereign operations. Key features of nonsovereign projects and initiatives are also discussed, where possible.

ABBREVIATIONS

ADB	Asian Development Bank
ASEAN	Association of Southeast Asian Nations
BTH	Beijing–Tianjin–Hebei
CCICED	China Council for International Cooperation on Environment and Development
CO_2	carbon dioxide
COP	Conference of the Parties
COVID-19	coronavirus disease 2019
CPS	country partnership strategy
DMC	developing member country
GDP	gross domestic product
GHG	greenhouse gas
$PM_{2.5}$	particulate matter 2.5
PRC	People's Republic of China
SMEs	small and medium-sized enterprises
TA	technical assistance
YREB	Yangtze River Economic Belt

EXECUTIVE SUMMARY

With the strong economic growth of the People's Republic of China (PRC) in the 2000s came significant impacts on the environment. These environmental challenges presented a unique opportunity to transition to green development. Throughout this decade-long journey, the Asian Development Bank (ADB) has partnered with the PRC to balance economic and social progress with environmental sustainability.

Starting in 2011 with the PRC's 12th Five-Year Plan, the government emphasized the greening of infrastructure, focused on environmentally sustainable rural development, and piloted green and low-carbon cities. Since then, through concerted efforts, the PRC has continuously promoted comprehensive, and innovative green development.

From 2011 to 2020, 92% of projects financed by ADB helped support environmentally sustainable development. In sovereign loan and grant investment projects, the share of the agriculture, natural resources, and rural development sector, which often includes green projects, has been constantly rising from 4% in 1991–2000 to 13% in 2001–2010 and further to 25% in 2011–2020. Another sector contributing to green development— the water and other urban infrastructure and services sector—has maintained its significant share, around 12%–16% over the 3 decades.

ADB's new country partnership strategy (CPS) for the PRC covering 2021–2025 supports government efforts to achieve high-quality, green development in line with the 14th Five-Year Plan. The new CPS focuses on three main pillars: environmentally sustainable development; climate change adaptation and mitigation; and aging society and health security. The four areas of collaboration that have emerged over the past decade will remain partnership priorities moving forward— enhancing environmental protection and ecological conservation, transforming the rural economy, developing green livable cities, and investing in climate change mitigation and adaptation. About $2.89 billion will be invested in green development projects in the Yangtze River Economic Belt (YREB) by 2023. ADB's support for the YREB focuses on ecosystem conservation, an integrated multimodal transport corridor, a green industrial transformation, institutional strengthening, and policy reforms to reduce the degradation of natural resources by pollution and overexploitation. ADB's flexible and long-term support for the YREB Development Plan is provided through a programmatic approach for lending and nonlending assistance, in close consultation with the government. Adoption of a green ecological corridor approach was a key innovation to coordinate public interventions for strengthening environmental management, ecological protection, and inclusive economic growth.

ADB's support for rural vitalization was formalized in 2018 with a memorandum of understanding with the National Development and Reform Commission and the Ministry of Finance. ADB support for rural vitalization has progressively evolved and shifted to more resilient and ecosystem-based approaches. Current and future ADB interventions emphasize environmental sustainability and resilience. This is supported by policy dialogue, demonstrative approaches, and green finance solutions that engage the private sector and use domestic and international cofinancing.

ADB's urban lending first focused on large cities and single sectors such as urban infrastructure and capacity development. In recent years, the partnership in urban development shifted to more holistic approaches. Several projects have aligned with ADB's Strategy 2030 and its operational priority on livable cities. These projects now contain a higher level of integration across sectors—aiming at green and inclusive, low-carbon, and climate-resilient urban development. Projects have also more directly supported green transformation and green economic development. ADB has been engaging the private sector in several urban projects, especially in green activities including water and wastewater, and solid waste management.

ADB has a long history of working with the PRC in addressing climate change. ADB and the PRC established a lending program, spanning 2015 to 2020, for six loans amounting to $2.1 billion to reduce air pollution in the greater Beijing–Tianjin–Hebei region. This assistance was later expanded to parts of Henan, Liaoning, Shandong, Shanxi provinces, and the Inner Mongolia Autonomous Region.

ADB's overall support in this area revolves around strengthening policies and building the capacities of institutions, developing custom-fit financing models, and advancing technology in key sectors and industries that can reduce both air pollutants and greenhouse gas emissions.

ADB supports low-carbon designs, climate-resilient infrastructure, and smart city platforms, including intelligent transport to stimulate low-carbon behaviors and practices.

ADB and the PRC will continue building on past joint ventures and new approaches to greening development for a more sustainable future.

ADB will continue aiming to create synergies in its green programs by designing projects bundling several green development objectives that reinforce each other. Incorporating digital platforms and other information technology applications knit the pieces together, and they are necessary in broadening communication and participation.

Green and sustainable finance is one key area in the coming years. The green business models and technologies, and green financial institutions developed in the PRC can be replicated within and outside the Asia and Pacific region for a bigger development impact.

Success in pursuing these initiatives rests largely on partnerships with various stakeholders to accelerate scaling-up and to seek more innovative approaches toward integrated, often spatially-based, high-quality green development. Effective collaboration will take the partnership through different pathways: fully testing the potential of new tools such as ecological redlining and eco-compensation; promoting the use of technological innovations; and bolstering further governance, policy reforms, and institutional and capacity strengthening. These pathways are needed to effect fundamental changes in the way environmental protection is managed in the PRC.

Piloting, dissemination, and replication of successful outcomes will be critical elements of the green development partnership between ADB and the PRC, bringing promising prospects to Asia and the Pacific and beyond.

Integrating flood and enviornmental risk management.
ADB supports building of ecological embankments along
rivers and lakes to manage floods and the ecosystem
(photo by Chongqing Project Management Office).

11TH FIVE-YEAR PLAN (2006–2010)
PROMOTING A RESOURCE-EFFICIENT, ENVIRONMENTALLY FRIENDLY SOCIETY

12TH FIVE-
STIMULATING G

Produced Agenda 21 White Paper to guide sustainable development.

Targets set for national forest cover at 18.2% and urban "green rate" at 35%, and passed the Environmental Assessment Law during the 10th Five-Year Plan (2001–2005).

Ministry of Environmental Protection created and upgraded as a full State Council Department.

Economic Stimulus Package with green components introduced.

Commitments established for the Aichi Biodiversity Targets (2011–2020) at the Convention on Biological Diversity Conference of Parties 10.

Environmental ministers' roundtable on environment and development strategies at Rio+20 held.

Environmental protection government strategy introduced at the 18th National Congress.

2006 — 2007 — 2008 — 2009 — 2010 — 2011 — 201

Established the State Council Leading Group on Environmental Protection.

Established the legal system of ecological and environmental protection through: *Environmental Protection Law (1979), Law on the Prevention and Control of Air Pollution (1987), and Law on the Prevention and Control of Solid Waste Pollution (1990).*

First National Climate Change Program launched.

Action Plan on Environment and Health (2007–2015) released.

Commitments set for low-carbon economy, forest carbon sinks, and 15% of renewable energy use by 2020 at the UN Climate Summit.

Law passed for circular economy for resource recovery and reuse.

Financial incentives provided for domestic solar power and new energy vehicles.

National Biodiversity Conservation Strategy and Action Plan (2011–2030) implemented.

Call for environme improvement by 2 announ

PLE'S REPUBLIC OF CHINA

PLAN (2011–2015)
VELOPMENT

13TH FIVE-YEAR PLAN (2016–2020)
FOSTERING ECOLOGICAL CIVILIZATION

War on Pollution on air,
ater, and soil announced.

v-type Urbanization Plan
(2014–2020) enhancing
ainability of cities issued.

Rural Vitalization Strategy adopted
to modernize rural economy,
address urban–rural disparities,
restore and improve ecosystems and
environment.

Ministry of Environmental
Protection becomes
Ministry of Ecology
and Environment with
climate change function.

Ministry of Natural
Resources established.

Law on Prevention and
Control of Soil Pollution
adopted.

PRC's Environmental
Protection Tax Law
takes effect.

2013 **2014** **2015** **2016** **2017** **2018** **2019** **2020**

Environmental Protection Law
updated with more effective
regulatory measures, better
enforcement, and broader
basis for eco-compensation.

National Plan for Sustainable
Development of Agriculture
(2015–2030) released.

National Plan for
Implementing
UN 2030 Agenda
for Sustainable
Development released.

PRC joins Minamata
Convention on Mercury.

Guidance for
Establishing Green
Financial Systems
issued.

PRC's National Park
System established.

Yangtze River
Economic Belt
planning framework
adopted.

National Ecological
Conservation
Red Line Program
established.

PRC aims to
reach national
"carbon
neutrality"
before 2060,
peaking carbon
by 2030.

Yangtze River
Protection Law
enacted.

Over the last 35 years, the partnership between ADB and the PRC has evolved and transformed to prioritize greener, more environmentally sustainable development, for a better balance between economic growth and environmental sustainability.

Protecting Hunan Miluo River from flooding. Structural measures with nature-based solutions help preserve Miluo River, which forms an important flood basin in the middle reach of the Yangtze River (photo by Pingjiang Project Management Office).

Improving sewerage systems. ADB is helping improve the sewage treatment system of townships and villages to lessen the rural pollution of the Xin'an River (photo by Mingyuan Fan).

INTRODUCTION

The Evolution

The People's Republic of China (PRC) is a development success story. In February 2021, the Government of the PRC officially announced it had eradicated absolute poverty. Citizens' well-being and quality of life significantly improved with the rapid economic growth beginning in the 1970s.[1] But this success came at significant cost to the environment.

Over the last 35 years, the partnership between the Asian Development Bank (ADB) and the PRC has evolved. It now focuses on supporting a range of green investments and interventions with a long-term programmatic approach, often encompassing large geographic areas. Over the past decade, the partnership has transformed to prioritize greener, more environmentally sustainable development. This transformation reflects the PRC's strategic, long-term commitment to achieving a better balance between economic growth and environmental sustainability.

The 1990s and early 2000s marked a shift away from more traditional, stand-alone transport and energy projects to environmental projects. These focused mainly in the water, sanitation, solid waste, and clean energy sectors. During this period, the PRC began to introduce key policy reforms in areas such as water tariffs, waste management, and renewable energy. The capacities of national, provincial, and local institutions were strengthened to support green development.

Based on strong environmental policy dialogue and enhancement of institutional capacities, the PRC and ADB laid the foundation for joint efforts to support green development. In the past 10 years, the PRC and ADB have worked together to address the challenges of environmental protection, biodiversity conservation, rural development, urbanization, and climate change. This partnership will continue to emphasize green development, which is fully aligned with the PRC's 14th Five-Year Plan, 2021–2025[2] and ADB's country partnership strategy (CPS), 2021–2025. ADB's new CPS 2021–2025 for the PRC focuses on "high-quality, green development" across three main pillars: environmentally sustainable development; climate change adaptation and mitigation; and aging society and health security.[3]

1   A. James. 2020. China Says It Has Met Its Deadline of Eliminating Poverty. *Wall Street Journal*. 23 November.
2   Huaxia. 2021. China's 14th Five-Year Plan Published in Booklet. *Xinhua*. 14 March.
3   ADB. 2021. *Country Partnership Strategy: People's Republic of China, 2021–2025—Toward High-Quality, Green Development*. Manila.

The Need for Greening

The greening of development in the PRC matters for many reasons. The PRC's response to the climate change impact, rural environmental degradation, and rapid urbanization will have profound implications for the country, the region, and the rest of the world toward realizing a greener and more sustainable future. The PRC is the largest greenhouse gas (GHG) emitter of carbon dioxide (CO_2), mainly from coal electricity generation.[4] Thus, its declaration in September 2020 to becoming carbon neutral by 2060 will create ripples of change to the region and the rest of the world.[5] The PRC's achievement will also be critical to collectively meet the targets of the Paris Agreement.

The PRC's progress in greening development will deepen awareness and knowledge—another public good—about best practices, innovations, and lessons learned in pursuing a more environmentally sustainable future. This knowledge is important not only for the PRC, but is essential for other developing member countries (DMCs) in Asia and the Pacific that are pursuing greener development pathways. The PRC's experience in designing innovative projects, introducing new policies, and strengthening institutional capacities can help offer models for DMCs to address national and regional environmental challenges.

The interdependence of human health and the natural environment has never been more starkly brought to light upon the onset of the coronavirus disease (COVID-19) pandemic. The call for a "green recovery" is resonating throughout the world to ensure that the return to growth and development after COVID-19 is sustainable and resilient.

The international community has recognized that green recovery efforts should address the global environmental concerns of climate change, ecosystem and biodiversity decline and pollution.[6]

The Partnership between the People's Republic of China and Asian Development Bank

The succeeding chapters describe how the partnership between the PRC and ADB has evolved over the past decade to focus on green development and presents future directions for continued collaboration. Chapter 2 summarizes how the 12th and 13th Five-Year Plans mainstreamed environmental considerations into the development agenda at the national, provincial, and local levels. Chapters 3 through 6 describe how the PRC and ADB have worked together over the past decade to put green development into practice in four areas:

(i)   Enhancing Environmental Protection and Ecological Conservation
(ii)   Transforming the Rural Economy
(iii)   Developing Green Livable Cities
(iv)   Investing in Climate Change Mitigation and Adaptation

Case studies are used to illustrate how projects were designed, financed, and implemented, and to showcase best practices and lessons learned. The final chapter provides a preview of the partnership between the PRC and ADB over the next 5 years under the 14th Five-Year Plan and the CPS in pursuing a greener and more sustainable future.

---

[4]   Global Carbon Atlas. CO_2 emissions. Accessed 20 July 2021. Available data as of 2019.

[5]   Ministry of Foreign Affairs of the People's Republic of China. 2020. President Xi Jinping's statement at the General Debate of the 75th Session of The United Nations General Assembly. 22 September.

[6]   United Nations Environment Programme. 2021. *Making Peace with Nature: A Scientific Blueprint to Tackle the Climate, Biodiversity and Pollution Emergencies*. Geneva.

Access to green spaces. ADB supported the rehabilitation of the Suzhou Creek, which improved sanitation services and provided greater access to parks and green spaces along the riverbanks (photo by ADB).

LAYING THE FOUNDATIONS

Recognizing the Green Imperative

Steady economic growth in the PRC during the early 2000s came with significant costs on the environment. Environmental problems became a national concern, especially urban air pollution. Water quality deteriorated significantly in major rivers passing through large cities. Groundwater contamination increased. Soil contamination by heavy metals and pollutants undermined the safety of local food production.

In 2007, the PRC became the world's largest GHG emitter, contributing to warming trends in the atmosphere and the oceans.[7] These climate change impacts exacerbated the likelihood and effects of extreme weather events, including droughts, floods, and intense storms, which were already undermining economic progress, and threatening food and water security in the PRC. The human health risks of deteriorating environmental conditions became a growing national concern.

By the late 2000s, substantial efforts were underway to reduce pollution and promote cleaner production under the concept of a circular, low-carbon economy. However, these efforts were mainly carried out within sectors. Economic growth continued to be the key measure of progress. The principal challenges included the economy's emphasis on heavy industry, continued reliance on coal, and high rates of urbanization with associated demands on infrastructure.

Solutions started to appear in new green technologies, sustainable production, and consumption initiatives, functional zoning and eco-compensation for land stewardship, and improved practices for biodiversity conservation. The first major efforts to tackle GHG emission intensity per unit of gross domestic product (GDP) were introduced in 2009, but it became clear that a more comprehensive and coordinated approach was required to address growing environmental challenges across the PRC.

Prioritizing Green Development

Beginning in 2011 with the 12th Five-Year Plan, green development became one of three key national priorities. The heavily polluted large coastal cities started to move the location of some industrial facilities, address solid waste management, and direct more investment toward environmental protection.

---

[7]   Climate Watch: Historical GHG Emissions. Accessed 2 July 2021.

In response to an alarming decline in biodiversity, the PRC initiated its 2011–2030 National Biodiversity Conservation Strategy and Action Plan (NBCSAP), operating in the context of the 2011–2020 Aichi Convention on Biological Diversity (CBD) goals and targets.[8] This action plan focused on improved management of the country's numerous nature reserves. These initiatives later became important for conservation and ecological restoration, including innovative methods for desertification control.

In 2014, the PRC's War on Pollution was declared. It was another important initiative during the 12th Five-Year Plan to address air, water, and soil pollution issues.[9] Greening of infrastructure became a higher priority along with an emphasis on environmentally sustainable rural development and the piloting development of green and low-carbon cities.

Toward the end of the 12th Five-Year Plan, many new green policies were introduced (Box A1), as well as ambitious goals for climate change, carbon intensity reductions, energy efficiency, and renewable energy targets. This set the stage for an even greener 13th Five-Year Plan.

Planning for Green Development

The PRC 13th Five-Year Plan (2016–2020) was the PRC's first full-fledged green national development plan. Green development became one of three key national priorities. Environmentally sustainable goals became very closely aligned with "high-quality" development. The PRC leadership expressed a new vision and philosophy of "Ecological Civilization." This builds on the foundation that the country could become prosperous based on values that included respect for nature, sustainability of natural resources, ecosystems valuation, a pollution-free environment, and measuring green growth rather than higher GDP rates.[10]

The 13th Five-Year Plan period is also when the rest of the world began moving more definitively toward green economy and green growth. During this time, the PRC joined the 2015 Paris Climate Change Agreement and participated in the United Nations (UN) 2030 Sustainable Development Goals. Due to this change of mindset acknowledged by the international community, the PRC was invited to host the Conference of the Parties (COP 15) to the CBD in 2021, at which a 10-year action plan for global biodiversity conservation is to be negotiated.

As part of the PRC's targets to achieve an ecological civilization, scaled-up efforts to protect biodiversity and natural resources were initiated under the 13th Five-Year Plan, including the demarcation of "ecological red lines" that protected biologically important areas. The PRC's NBCSAP requires that biodiversity is "effectively protected" by 2030 and lists 35 priority regions and actions for conservation. Over half of these are in the Yangtze and Yellow river basins. Several transformative changes during the 13th Five-Year Plan, including the PRC's cooperation on international efforts to address climate change and environmental sustainability, served to reaffirm the PRC's long-term commitment to both national and global green development.

The 14th Five-Year Plan (2021–2025) saw a move away from economic growth and restructuring. It focuses on the sustainability

---

[8]   Position Paper of the PRC for the UN Summit on Biodiversity. 2020. *Building a Shared Future for All Life on Earth: China in Action*. 21 September.

[9]   The National People's Congress of the People's Republic of China. 2014. China Declares War Against Pollution. *Xinhua*. 21 September.

[10]   A. Hanson. 2019. Ecological Civilization in the People's Republic of China: Values, Action, and Future Needs. *East Asia Working Paper* No. 21. ADB. Manila.

of growth and the quality of life.[11] The plan paves the way for the objective to peak CO_2 emissions before 2030 and reach carbon neutrality before 2060 and sets green development targets for energy and carbon intensity, air quality, surface water quality, and forest coverage.[12] As a nonbinding indicator, the proportion of nonfossil fuels in primary energy consumption is set at 20% from 15% in the previous plan. The plan promotes low-carbon development and the circular economy with new approaches to transport, energy production, and waste management policies.

Partnering with the Asian Development Bank

ADB's country strategies and programs through the 1990s and 2000s reflected environmental protection and sustainability as a priority area, supporting capacity building for environmental assessment, and institutional improvement including legal, regulatory, and policy frameworks; addressing air, water, and soil pollution; promoting energy efficiency, integrated urban development, and sustainable rural ecosystems; and incorporating environmental considerations in ADB-funded projects.

ADB's CPS 2011–2015 saw a major shift toward green development, which mainstreamed environmentally sustainable growth. During this CPS period, ADB operations were prepared and assessed with the environment in mind, to promote further greening of the portfolio and to mainstream climate change considerations into investments.

Under ADB's CPS 2016–2020, managing climate change and the environment was one of its three strategic priorities. Integrated river basin management, especially under the Yangtze River Economic Belt Program, was a principal area of partnership for environmental protection and ecological conservation.

This strategic transformation is evident in ADB programs and portfolio in the PRC. In sovereign loan and grant investment projects, the share of the agriculture, natural resources and rural development sector, which often includes green projects has been constantly rising from 4% in 1991–2000 to 13% in 2001–2010 and further to 25% in 2011–2020 (Figure 2.1).

Figure 2.1: Operational Shift toward Green Support, 1991–2020

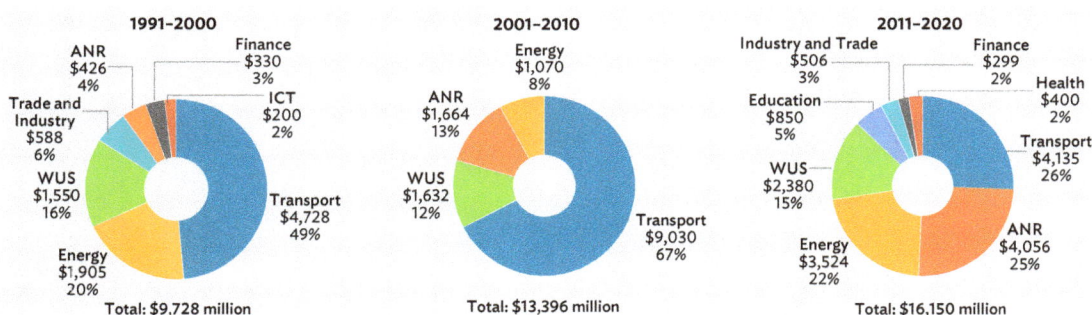

1991–2000
- ANR $426 — 4%
- Finance $330 — 3%
- ICT $200 — 2%
- Trade and Industry $588 — 6%
- WUS $1,550 — 16%
- Energy $1,905 — 20%
- Transport $4,728 — 49%
- Total: $9,728 million

2001–2010
- Energy $1,070 — 8%
- ANR $1,664 — 13%
- WUS $1,632 — 12%
- Transport $9,030 — 67%
- Total: $13,396 million

2011–2020
- Industry and Trade $506 — 3%
- Finance $299 — 2%
- Health $400 — 2%
- Education $850 — 5%
- WUS $2,380 — 15%
- Transport $4,135 — 26%
- Energy $3,524 — 22%
- ANR $4,056 — 25%
- Total: $16,150 million

ANR= agriculture and natural resources, ICT = information and communication technology, WUS= water and other urban infrastructure and services.
Note: Includes sovereign loans and investment grants. Percentages may not total 100% due to rounding.
Source: Asian Development Bank loan, technical assistance, grant, and equity approvals database.

[11] ADB. 2021. *PRCM Observations and Suggestions: The 14th Five-Year Plan of the People's Republic of China—Fostering High-Quality Development*. Manila.

[12] The PRC's water quality classification system is based on the purpose of use and protection target. Grade III includes water source protection area for centralized drinking water supply, sanctuaries for common species of fish, and swimming zones.

Another sector contributing to green development—the water and other urban infrastructure and services sector—has maintained its significant share around 12%–16% over the 3 decades.

Other infrastructure sectors have been increasingly supporting energy efficiency, renewable energy, and environmentally sustainable transportation. Nonsovereign operations have also been more focused on green development. Sector priority shifted toward more environmentally sustainable development.

From 2011 to 2020, green development projects became the norm rather than the exception—92% of projects financed by ADB directly or indirectly support environmentally sustainable development (Figure 2.2 and Appendix Table A1). These operations promote (i) environmental protection and ecological conservation; (ii) rural economy; (iii) green livable cities; and (iv) climate change mitigation and adaptation (Figure 2.3). There was a visible shift.

In 2011–2015, ADB's loan portfolio supporting environmentally sustainable development investments remained under the traditional infrastructure sectors such as transport, agriculture and natural resources, energy, and water and other urban infrastructure and services. Then in 2016–2020, lending operations supporting environmentally sustainable growth expanded to include projects in the non-infrastructure sectors such as health, finance, education, and industry and trade (Appendix Figure A1).

ADB has been collaborating for greener support with the PRC's other development partners in areas such as natural resources management and biodiversity conservation, pollution reduction (air, water, soil, and marine), strengthening climate resilience (adaptation), and promoting low-carbon development (mitigation). For ADB-funded

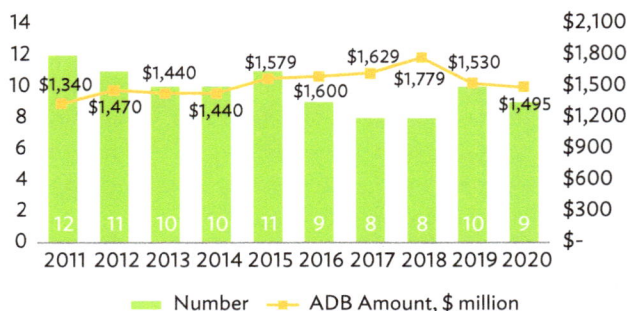

Figure 2.2: Asian Development Bank Approvals Supporting Green Development in the People's Republic of China by Year, 2011-2020

ADB = Asian Development Bank.

Note: Includes 98 sovereign loans approved during 2011–2020 that contribute to ADB's environmentally sustainable growth strategic agenda.

Sources: ADB loan, technical assistance, grant, and equity approvals database; ADB eOperations; and ADB East Asia Department.

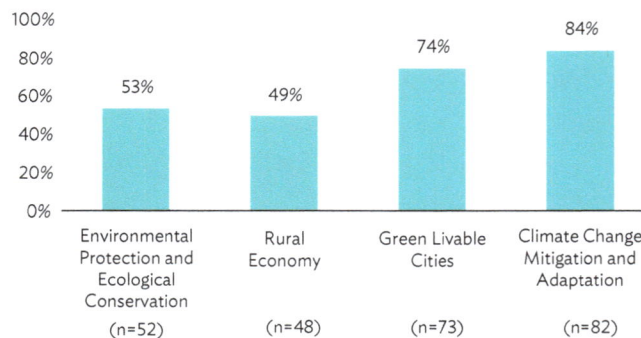

Figure 2.3: Asian Development Bank Approvals Supporting Green Development in the People's Republic of China by Theme, 2011–2020 (%)

ADB = Asian Development Bank, n= number.

Note: Includes 98 sovereign loans approved during 2011–2020 that contribute to ADB's environmentally sustainable growth strategic agenda. Total percent does not add up to a 100% because one operation may contribute to more than one theme.

Sources: ADB loan, technical assistance, grant, and equity approvals database; ADB eOperations.

green projects committed between 2018 and 2020, every $1 from ADB sovereign lending brought in another dollar of cofinancing from key development partners.

The future. ADB's new CPS for the PRC covering 2021–2025 will support government efforts to achieve high-quality, green development in line with the 14th Five-Year Plan. The four areas of collaboration that have emerged over the past decade will remain partnership priorities moving forward—enhancing environmental protection and ecological conservation, transforming the rural economy, developing green livable cities, and investing in climate change mitigation and adaptation.

From 2011, ADB's operations in the PRC saw a major shift toward green development, which mainstreamed environmentally sustainable growth to promote further greening of the portfolio.

The following chapters show how these four partnership priorities have evolved in terms of enabling policies and plans, as well as through investment projects, technical assistance, and knowledge and capacity building support. The lessons learned that guide the future, from piloting new initiatives, introducing policy reforms, and replicating good practices on green development are also highlighted.

Protecting water resources. ADB is strengthening wastewater management and pollution control to help improve water resources management in the Sayu River basin in Yunnan, PRC (photo by Zhaotong Yudong Reservoir Management Bureau).

3

Cleaning Xin'an River. ADB is helping clean up the Xin'an River to improve public health and ensure a sustainable and reliable source of drinking water for years to come (photo by Mingyuan Fan).

ENHANCING ENVIRONMENTAL PROTECTION AND ECOLOGICAL CONSERVATION

The Greening of the Yangtze River Basin

The Yangtze River is the longest river in the PRC and Asia and the third longest river in the world.[13] Known as the "Mother River" of the PRC, the Yangtze River provides about 40% of the country's freshwater, and accounts for 20% of its total wetland areas.[14] About 600 million people use its resources for drinking water. The basin is also one of the world's most biologically diverse eco-regions. The river-lake linkages are climatically, geographically, and geomorphologically diverse, and hydrologically complex.[15] Also called the "golden waterway," it is the key transport corridor between interior and coastal regions. The region it covers is the economic powerhouse of the PRC, generating about 40% of the national GDP. Protecting the Yangtze River ecosystem is of vital importance for socioeconomic development and ecological security.

Overexploitation of its resources has put the health of the ecological system under immense pressure. Pollution of the river comes from several sources—industrial wastes, domestic sewage, unsustainable agricultural production systems, and inadequate land management practices. Weak coordination among the basin's nine provinces has also been one of the main causes of stark regional disparities. Limited financial opportunities were further exacerbated by a changing climate and frequent occurrence of extreme weather events.

Replacing "big development with big protection" for the Yangtze River Basin was announced in 2016.[16] To achieve this goal, the government formulated the Yangtze River Economic Belt (YREB) Development Plan (2016–2030). The plan introduced the geographic concept of a "belt" that encompasses economic, social, and ecological linkages. The centerpiece of the plan is balanced development between socioeconomic growth and environmental sustainability, with restoration and conservation

---

[13] United States Geological Survey. *Rivers of the World: World's Longest Rivers.*

[14] S. Groff. 2018. Supporting PRC's "Mother River" Will Help Achieve Ecological Civilization. *Asian Development Blog.* 9 February.

[15] WWF China. 2020. *Living Yangtze Report 2020 Summary.*

[16] This political will was confirmed at the opening session of the National People's Congress in Beijing on 22 May 2020, where the Premier Minister Li Keqiang highlighted the need for advancing the well-coordinated environmental conservation in the YREB.

of the Yangzte ecosystems. The plan also aims to address regional inequalities along the river from eastern to western provinces. To achieve these ambitious goals, the government has adopted an integrated model to improve ecological rehabilitation, natural resources management, and advanced agricultural development, striving to coordinate interventions across regional and provincial governments.

Introducing Eco-Compensation

The PRC has been a pioneer in applying the concept of "ecological compensation" to protect the environment and livelihoods given competing demands for natural resources. As defined by the PRC's National Development and Reform Commission (NDRC), eco-compensation is built around the concept of providing rewards for protecting ecosystems and natural resources, compensation for environmental damage, and fees charged for those who pollute the environment. In the YREB, eco-compensation has become a key policy incentive mechanism that factors in the value of the environment and natural resources.[17] Provincial, municipal, and county governments have carried out pilot demonstrations on eco-compensation since the early 2000s. Eco-compensation schemes have enabled ecological outcomes and regional integration and helped reduce poverty.

Eco-compensation can also promote cooperation across provinces to improve water quality across the river's upstream and downstream areas. In February 2018, a special meeting on the Yangtze river

eco-compensation was held in Chongqing. Provinces and cities along the Yangtze River Basin that implemented transboundary watershed eco-compensation schemes were rewarded. However, during this period, private sector participation in eco-compensation schemes had been limited. Despite the great efforts in leveraging economic incentives, eco-compensation investments continue to be driven by the public sector and function mainly as budget transfer support mechanisms from the central government to the provinces.

The Yangtze River Protection Law, which entered into force in March 2021, is the first legislation in the PRC to tackle complex environmental challenges at river basin scale.[18] By promoting an ecosystem-based approach, the law will help to reinforce the role of river basin protection institutions. This includes the Yangtze River Basin Ecological and Environmental Supervision and Administration Agency. As the central office of the Yangtze River Basin Coordination Body, the agency will help protect the ecology and environment through a supervisory role for pollution prevention and ecological protection. The unified approach helped to achieve better coordination, and covered environmental assessment, monitoring, data access, and evaluation functions. The law aims to strengthen land-use and spatial management. It introduced the national concept of "Ecological Conservation Red Lines"[19] at basin level, which will implement integrated supervision, ecological monitoring, evaluation, and law enforcement. The law also provides for an increased role of the public, so that citizens and the community help in decision-making processes.

---

[17] ADB. 2016. *Toward a National Eco-Compensation Regulation in the People's Republic of China.* Manila.

[18] The Yangtze River Protection Law of the People's Republic of China.

[19] Around the Yangtze River delta, 28,995 square kilometers of land is set aside for protection, including Yangtze River shorelines, important wetland forest, and grassland.

Partnering to Support the Yangtze River Economic Belt Program: The Green Ecological Corridor Approach

ADB's role in supporting efforts to prepare the YREB development plan has been pivotal. Adoption of a green ecological corridor approach was a key innovation to coordinate public interventions for strengthening environmental management, ecological protection, and inclusive economic growth. ADB supported the plan for the YREB development across four areas:

(i) **Institutional strengthening, governance, and policy reforms.** Policy and regulatory reforms for integrated river basin management across local governments, environmental management, pollution control, ecosystem restoration, biodiversity protection, and climate change adaptation and mitigation (i.e., river basin laws, eco-compensation regulation and transprovincial

management, national climate strategy, natural capital accounting and gross ecosystem product). An example is provided in Box 3.1.

(ii) **Natural resources management, ecosystem restoration, biodiversity conservation, and sustainable management of water resources.** Protection, management, and restoration of nature and ecosystems with use of green infrastructure and nature-based solutions (i.e., prevention and control of pollution, integrated flood risk management, watershed protection, eco-compensation demonstration, nonpoint source pollution control, wetland restoration and biodiversity conservation). A project example is provided in Box 3.2.

(iii) **Green development and inclusive low-carbon transformation.** Private sector engagement for green development and industrial transformation (i.e., green fund mechanisms for private sector

Box 3.1: Managing Water Resources in the Yunnan Sayu River Basin

Water resources in the Sayu River Basin. ADB is using sustainable funding mechanisms to manage water pollution in the Sayu River Basin (photo by Zhaotong Yudong Reservoir Management Bureau).

The Yunnan Sayu River Basin Eco-Compensation Demonstration Project will promote regional public goods by reversing the negative impacts of pollution. More than 15,000 households in the Sayu River Basin will benefit from connected wastewater facilities through innovative integrated water pollution management mechanisms. Pilot eco-villages will be developed, and new approaches such as low-emission agriculture will be introduced as part of eco-compensation measures to manage water resources in the Yunnan Sayu River.

The project design incorporates lessons from previous ADB-financed projects in promoting integrated water resources management, including capacity development in project implementation and effective monitoring and evaluation.

Source: ADB. 2020. *Report and Recommendation of the President to the Board of Directors: Proposed Loan to the People's Republic of China for the Yunnan Sayu River Basin Rural Water Pollution Management and Eco-Compensation Demonstration Project.* Manila.

Box 3.2: Mitigating Flood Risks in the Chongqing Longxi River Basin

Integrated approach to reduce flood risks. ADB is developing local capacity to reduce flood and environmental risk in the Longxi River watershed (photo by Chongqing Project Management Office).

The Asian Development Bank-financed water-based natural resources project will help mitigate flood and environmental risk in the Longxi River watershed. The integrated approach at river basin scale prioritizes upstream–downstream and urban–rural linkages, promotes nature-based solutions, integrates flood and environmental risk management into regional planning, and develops capacity for project implementation.

When the coronavirus disease (COVID-19) broke out, this integrated approach enabled the project implementation team to quickly adapt to the changing needs. Multisector emergency response resources were mobilized, such as facilities, equipment, and specialized emergency skillsets for pandemic management. These coordinated efforts further expanded the flood and environmental risk management to include public health emergency management, laying foundation for cooperation beyond the water and environment sectors.

Source: ADB. 2018. *Report and Recommendation of the President to the Board of Directors: Proposed Loan to the People's Republic of China for the Chongqing Longxi River Basin Integrated Flood and Environmental Risk Management Project*. Manila.

engagement, economic incentives for pollution control management, integrated urban development for livable cities and low-carbon urban design). A project example is provided in Box 3.3.

(iv) **Sustainable agriculture and rural–urban integration.** Coherent and integrated interventions to improve agricultural development, tackle environmental degradation, address poverty and interprovincial disparities in alignment with the the Rural Vitalization Strategy (i.e., sustainable land-use and agricultural practices, sustainable livestock and waste management, provision of alternative livelihoods and sustainable value chains). A project example is provided in Box 3.1.

ADB supported the YREB through a programmatic and basin-wide approach. The support focuses on ecosystem conservation, integrated multimodal transport corridor, green industrial transformation, institutional strengthening, and policy reforms to reduce pollution and degradation of natural resources.

By 2023, ADB would have invested about $2.89 billion on green development projects in the YREB since 2015. ADB approved 12 projects from 2016 to 2020 to demonstrate multisector approaches, innovative technologies, institutional strengthening, and policy reforms in the YREB (Figure 3.1, Appendix Table A2). Technical assistance projects strengthened planning, coordination, integration, and implementation of holistic development approaches for YREB transformation.[20]

---

[20] ADB. 2019. *Technical Assistance Completion Report: Strengthening Provincial Planning and Implementation for the Yangtze River Economic Belt*. Manila.

Figure 3.1: The Asian Development Bank's Yangtze River Economic Belt Framework Approach

THE YANGTZE RIVER ECONOMIC BELT (YREB) COVERS:

9 PROVINCES

2 MUNICIPALITIES

$2.89 BILLION

IN ADB LOANS AND TECHNICAL ASSISTANCE FOR THE MIDDLE AND UPPER REACHES OF THE YREB FROM 2015–2023

AREAS OF INTERVENTION

| Institutional strengthening, governance, and policy reforms | Natural resources management, ecosystem restoration, biodiversity conservation, and sustainable management of water resources | Green development and inclusive low-carbon transformation | Sustainable agriculture and rural–urban integration |

ADB = Asian Development Bank, CO_2 = carbon dioxide.
Source: East Asia Department, ADB.

Box 3.3: Protecting the Anhui Huangshan Xin'an River through Green Incentives

Tea garden in Anhui Huangshan. Tea farmers who adopt sustainable farming practices are provided financial incentives (photo by Mingyuan Fan).

Two green financing mechanisms helped change farmers' behavior toward adoption of sustainable farming practices. As an example of eco-compensation, a "green incentive mechanism" provided direct incentive payments, through cash grants, to tea farmers to encourage environmentally sustainable farming. New livelihoods based on ecotourism, and businesses in ecological agriculture, environmental improvements, and other green enterprises can be supported through a "green investment fund."

The project will also use stormwater management modeling for analysis, planning and design for low impact site development. Local infrastructures will be improved by building drainage (sponge) ditches to help control pollution. Once completed, this project will provide 95% of the urban population access to wastewater treatment services.

Source: ADB. 2019. *Report and Recommendation of the President to the Board of Directors: Proposed Loan to the People's Republic of China for the Anhui Huangshan Xin'an River Ecological Protection and Green Development Project*. Manila.

ADB will provide policy advice and knowledge support to the first national regulation on eco-compensation regulation for 2022. The regulation builds on over 20 years of provincial and regional experience in piloting eco-compensation for ecosystem restoration, reforestation, water pollution control, sustainable land-use management, and regenerative agriculture. The three project examples highlight some of the key elements of ADB support to YREB key areas of intervention (Boxes 3.1, 3.2, and 3.3).

Guiding the Future

Key lessons can be drawn from the YREB experience that could guide other river basin management efforts in the region.

First, a national strategy, and efficient coordination from the national government are important in addressing interprovincial issues related to the environment, transport systems, and economic disparity. The YREB Development Plan provided the necessary framework for institutional coordination, especially among provinces and local governments.

Second, flexible and long-term support for the YREB Development Plan was made possible through a programmatic approach for lending and nonlending assistance, in close consultation with the Government of the PRC.

Third, balancing the national government's agenda and provincial government needs is paramount. Planning tools and integrated approaches can help provincial governments align their project proposals to the thematic areas and principles of the national YREB Plan.

Fourth, environmental protection should be linked with the provision of livelihood opportunities and poverty reduction in rural areas. Directly supporting livelihood opportunities, including technical and vocational education, improves the willingness of local farmers to comply with environmental regulations.

Fifth, coordination is vital in monitoring progress and assessing social and environmental concerns. There is a growing need for improved cooperation and effective communication among sectors, locations (upstream–downstream), and governments at various levels with the active engagement of stakeholders and communities.

Improving agricultural practices. ADB is supporting circular agricultural practices and helping produce organic fertilizer to enhance soil fertility in Pingjiang County in Hunan, PRC (photo by Pingjiang Project Management Office).

TRANSFORMING THE RURAL ECONOMY

Pursuing a Holistic Approach

Rural areas in the PRC are faced with multiple challenges. Aging population, inadequate finance, limited health and education services, outdated infrastructure and digital technology, labor outmigration,[21] lack of skilled workers, inadequate local governance and capacity, and collective land ownership issues are only some of the factors that limit rural development. Major rural development reforms have been implemented since the late 1970s. These focused on increasing food production and maintaining grain self-sufficiency through intensification of agricultural production. They led to a remarkable expansion of agricultural production and a drastic reduction in rural poverty—but at the expenses of nature in the rural environment.

Traditional approaches to rural development have contributed to significant environmental degradation. Unsustainable agricultural practices, such as intensive use of fertilizers, pesticides, and herbicides; deforestation; and intensive livestock production caused soil degradation, river and lake pollution, and decreased biodiversity. In addition to these pressures, weak institutional and governance arrangements have resulted in overlapping mandates and poor coordination across ministries, implementing agencies, provinces, and local governments. These factors, compounded with increasing climate change impacts, have continually threatened the sustainable use of natural resources.

To tackle many of these challenges, the Government of the PRC launched the Rural Vitalization Strategy as a multisector approach for integrated rural development. This concept was first presented in 2017 with a comprehensive focus based on ecological civilization and reduction of rural–urban income disparities.[22]

A rural vitalization plan followed, propelled by the State Council, to drive the effort of supporting low-income rural populations, addressing regional disparities, and reducing the size of vulnerable population. To achieve this target, the Ministry of Agriculture was reorganized into the Ministry of Agricultural and Rural Affairs in 2018. Promoting rural vitalization has gone hand in hand with consolidating and expanding the achievements in poverty reduction. In February 2021, the Government of the PRC officially announced it had eradicated absolute poverty after a long and challenging 40-year journey.[23] Despite rising rural per capita income, rural–urban income disparities persist. Reducing

---

[21] Eighty million rural residents are expected to move into cities, reducing the country's rural population to about 35% of its total.

[22] X. Mu. 2017. China Outlines Roadmap for Rural Vitalization. *Xinhua*. 29 December.

[23] Huaxia. 2021. Poverty Alleviation: China's Experience and Contribution. *Xinhua*. 6 April.

Box 4.1: Rural Vitalization

Rural vitalization promotes rural green and low-carbon development to ensure national food security and provision of high-quality ecological agricultural products, while protecting the rural eco-environment, reducing rural–urban disparities, attaining a "moderately prosperous society." This is achieved through multisector approaches, use of transformative technologies, and inclusive financial innovation combined with enhanced institutions, policy coordination, and market connectivity.

Source: Asian Development Bank and the Government of the People's Republic of China. 2018. *Memorandum of Understanding Between the National Development Reform Commission and the Ministry of Finance, People's Republic of China and the Asian Development Bank on Support for Rural Vitalization in the People's Republic of China.* Beijing.

relative poverty further and narrowing the rural–urban income disparity remains major policy issues. The PRC's post-2020 poverty reduction strategy will treat poverty as multidimensional and reinvigorate rural development. Therefore, a more holistic strategy is needed to reduce vulnerability and achieve sustainable development.

Operationalizing Rural Vitalization

ADB's support for rural vitalization was formalized only in 2018 with the signing of a memorandum of understanding (MOU) with the NDRC and the Ministry of Finance, which defined an overall assistance package of up to $6 billion (during 2018–2022) with financing support from ADB and other development partners.[24]

ADB's first generation of projects aimed to promote poverty reduction and rural development toward green and inclusive growth. To achieve this goal, ADB loans and technical assistance focused on solid waste and rural waste management, agricultural pollution control, agriculture modernization, improved provision of public services in rural areas (including health, education, elderly care, labor training), and integrated rural–urban development.

ADB support for rural vitalization has progressively evolved and shifted to more resilient and ecosystem-based approaches. Current and future ADB interventions emphasize environmental sustainability and resilience, supported by policy dialogue, demonstrative approaches, and green finance solutions leveraging private sector engagement and domestic and international cofinancing. Rural economies need to support productive, competitive, and innovative enterprises, and to provide sound social services toward decreasing inequalities and regional disparities.

ADB's second generation of interventions to support rural vitalization in the PRC focus on achieving rural green development through four core areas: (i) transforming food systems, (ii) enhancing market connectivity and digitalization, (iii) protecting the rural environment, and (iv) strengthening rural economies. Figure 4.1 suggests that there can be iterations among these core areas, with the potential for continuous improvement.

The Rural Vitalization Strategy is a multisector approach for integrated rural development, with a comprehensive focus on ecological civilization and reduction of rural–urban income disparities.

---

[24]   Asian Development Bank and the Government of the People's Republic of China. 2018. *Memorandum of Understanding Between the National Development Reform Commission and the Ministry of Finance, People's Republic of China and the Asian Development Bank on Support for Rural Vitalization in the People's Republic of China.* Beijing.

Figure 4.1: Four Core Areas for the People's Republic of China and the Asian Development Bank's Rural Vitalization Partnership

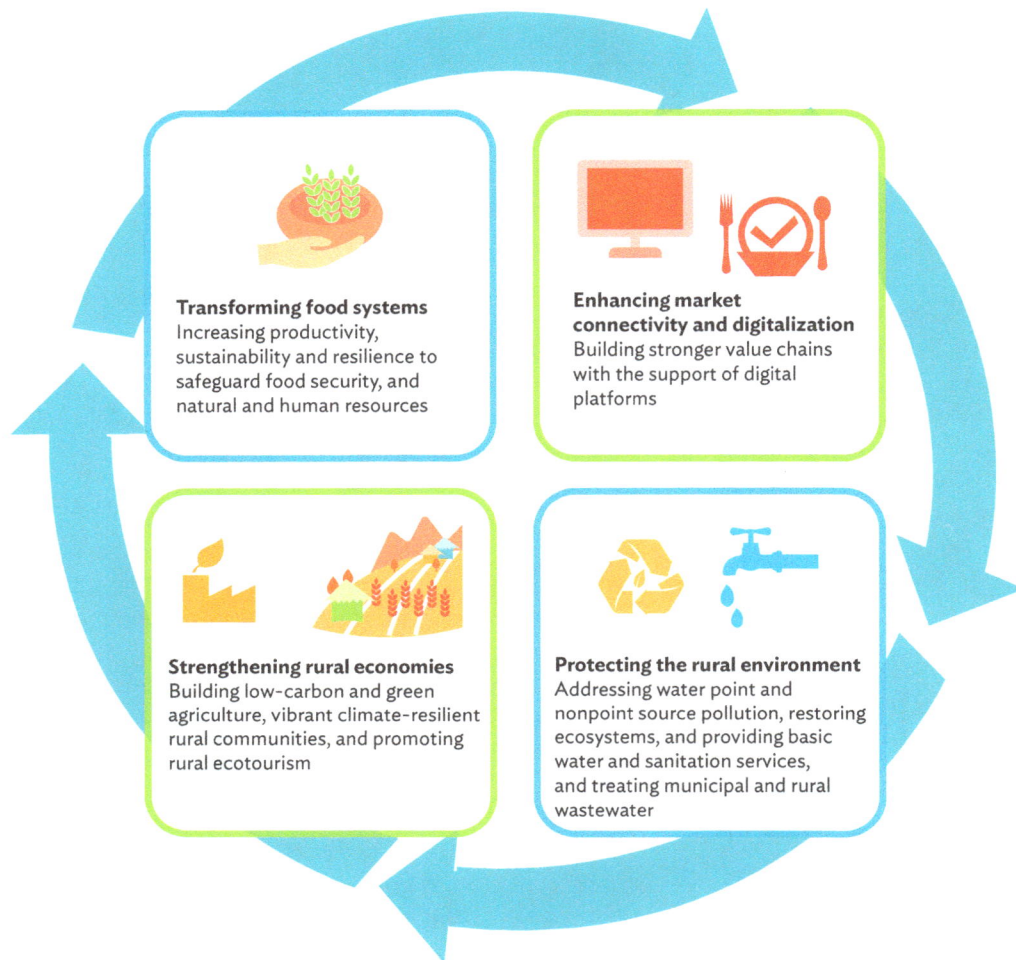

Transforming food systems
Increasing productivity, sustainability and resilience to safeguard food security, and natural and human resources

Enhancing market connectivity and digitalization
Building stronger value chains with the support of digital platforms

Strengthening rural economies
Building low-carbon and green agriculture, vibrant climate-resilient rural communities, and promoting rural ecotourism

Protecting the rural environment
Addressing water point and nonpoint source pollution, restoring ecosystems, and providing basic water and sanitation services, and treating municipal and rural wastewater

Source: S. Cardascia and S. Robertson. Forthcoming. Scaling Natural Capital Investments in the Yellow River Ecological Corridor. Manila.

Demonstrating Proof of Concept

One of the first large-scale and integrated approaches to greening the rural economy can be traced back to 2012 when ADB supported the Government of the PRC's priority to address the nation's food security problems. The Comprehensive Agricultural Development Project included five provinces—Anhui, Heilongjiang, Henan, Jilin, and Yunnan—and one autonomous region, Ningxia Hui. It combined investments in irrigation and drainage infrastructure with capacity building for farmers to adopt more advanced technologies and connect to markets. By 2018, the project improved 92,227 hectares (ha) of land to better resist drought and enhanced the skills of nearly 200,000 farmers in adopting modern agricultural technologies, significantly contributing to food security and growth in farm income.[25]

Consistent with the MOU on rural vitalization, ADB approved in 2018 an agriculture and natural resources project to promote sustainable and modern farming systems in five provinces and one municipality in the upper and middle reaches of the Yangtze River Basin (Box 4.2).

ADB continues to support environmental improvement and green development in the YREB. In Hunan province, for example, a rural environmental improvement project will improve rural waste and sanitation management, develop local special agricultural products, and integrate the urban and rural industries in Xiangxi Tujia and Miao Autonomous Prefecture, one of the poorest regions in the PRC.[26]

Linking basin-wide approaches such as the YREB with thematic countrywide programs like the rural vitalization will be critical to combine environmental protection efforts with provision of livelihood opportunities for rural green development. Creating synergies across diverse geographic areas is one of the main challenges that the new generation of ADB rural vitalization investment projects will need to address.

Guiding the Future

Some of the lessons in fostering rural vitalization that could guide future interventions include the following.

First, private sector enterprises, farmers, local government agencies, and other project stakeholders need to be highly involved.

Box 4.2: Promoting Climate-Smart Agriculture Practices in the Yangtze River Basin

Promoting sustainable farming in the Yangtze River Basin. Integrated pest management and organic fertilizers are used to reduce pollution and optimize crop nutrient requirements (photo by ADB).

Approved by the Asian Development Bank in 2018, this project promotes sustainable and modern farming systems, reduces pollution caused by diffused sources and environmental degradation, and strengthens institutional capacity for coordinated natural resources management in the upper and middle reaches of the Yangtze River Basin. The project is introducing modern farming practices to enhance productivity such as climate-smart farming practices, including water-saving measures for drip irrigation and water capture technology (adaptation), and balanced fertilizer use and reforestation (mitigation) to enhance climate resilience. The project will also use integrated pest management and organic fertilizers to reduce pollution and optimize crop nutrient requirements. About 115,000 hectares of farmland will be improved, which will support rural livelihoods and benefit 1.787 million people in poor and less developed rural areas of the project. The project will work at the watershed level to maximize the benefits of environmental rehabilitation and improved natural resources management.

Sources: ADB. 2012. *Report and Recommendation of the President to the Board of Directors: Proposed Loan to the People's Republic of China for the Comprehensive Agricultural Development Project*. Manila; ADB. 2018. *Report and Recommendation of the President to the Board of Directors: Proposed Loan to the People's Republic of China for the Yangtze River Green Ecological Corridor Comprehensive Agriculture Development Project*. Manila; and United States Environmental Protection Agency. Basic Information about Nonpoint Source (NPS) Pollution.

[25] ADB. 2021. *Completion Report: Comprehensive Agricultural Development Project in the People's Republic of China*. Manila.
[26] ADB. 2020. *Report and Recommendation of the President to the Board of Directors: Proposed Loan to the People's Republic of China for the Hunan Xiangxi Rural Environmental Improvement and Green Development Project*. Manila.

Box 4.3: Using Nature-Based Solutions to Improve the Living Environment along the Hunan Miluo River in Hunan Province

Safeguarding ecological services along the Hunan Miluo River. ADB is supporting biodiversity conservation, groundwater recharge, and maintaining soil fertility (photo by Pingjiang Project Management Office).

The project, approved in 2020, will improve the rural living environment for at least 500,000 people living in Pingjiang County. Health risks will be reduced through improved drinking water and sewage treatment. Nature-based solutions will be introduced, such as targeted restoration of wetland areas to protect and enhance the ecological services, and the establishment of county-wide early warning systems to reduce flood vulnerability for local communities. The project is also one of the first to pilot innovative green procurement practices and raise environmental awareness by leveraging youth engagement in schools.

Source: ADB. 2020. *Report and Recommendation of the President to the Board of Directors: Proposed Loan to the People's Republic of China for the Hunan Miluo River Disaster Risk Management and Comprehensive Environment Improvement Project.* Manila.

Their participation and involvement are crucial factors in the effective design and implementation of projects, which leads to sustaining the projects' future operations and maintenance.

Second—as we have learned from the COVID-19 pandemic—balancing the health of people, animals, and the environment is essential for sustainable rural development and calls for a cross-sector and a holistic food chain approach beyond agriculture and environmental sustainability.

Third, the PRC's ambitious target of achieving carbon neutrality by 2060 requires that rural economies be an integral part of the solution. Establishing a circular economy, using nature as a carbon sink, can contribute to the carbon neutrality agenda. From a financial sustainability perspective, agriculture and forestry restoration activities should be part of carbon credit markets to provide incentives for conservation practices.

Fourth, the rural vitalization agenda needs to embrace evolving opportunities for digital

technologies more broadly. At the production stage, big data analytics, Internet of Things, and sensors will help farmers' decision making through accurate, timely, and location-specific data on price, weather, and agronomic information. At the marketing stage, digital technologies and blockchain technologies can significantly reduce transaction costs and improve product traceability integrity, and compliance with sanitary and phytosanitary requirements. An example is the Gansu Internet-Plus Agriculture Development Project that applies network-connected information technology along the entire value chain from production to marketing.[27]

Lastly, institutions should be strengthened. Developing more comprehensive agri-environmental policies that include voluntary and mandatory environmental targets requires this. Improving local capacities is necessary for monitoring of environmental performance, enforcing environmental regulations, setting the level of eco-compensation programs, and managing local certification schemes.

---

[27] ADB. 2019. *Report and Recommendation of the President to the Board of Directors: Proposed Loan to the People's Republic of China for the Gansu Internet-Plus Agriculture Development Project.* Manila.

ADB support to rural vitalization focuses on transforming food systems, enhancing market connectivity and digitalization, protecting the rural environment, and strengthening rural economies.

Protecting villages from flooding. Building nature-based buffers along riverbanks and riverbank reinforcements in towns along the Miluo River helps protect the villages from flooding (photo by Au Shion Yee).

5

Green cities. Urban residents enjoy healthy lifestyle options (photo by ADB).

DEVELOPING GREEN LIVABLE CITIES

Living in Green and Sustainable Cities

With more than 60% of people in the PRC living in urban areas, green and sustainable urbanization has now become critical. Earlier, rapid urbanization from early 1980s until the 2000s was modeled based on wide roads and large blocks with separation of land-use and environments that are unfriendly to pedestrians. Now mayors, planners, and the growing middle class and urban residents have high expectations for "clean, green, and healthy" lifestyle options.

As early as the 1980s, the PRC started to adopt policies and programs that promote green cities and environment improvement. In 2009, the innovative and comprehensive "circular economy" law became effective. It showed the PRC's aim on resource and environmental conservation, improved recycling, and solid waste management, reducing industrial waste and pollution.

In 2013, the National Climate Change Adaptation Strategy emphasized risk management and improving disaster response systems in cities. It was designed to enhance human health, infrastructure, and other private and public investments. In 2016, the Urban Climate Change

Adaptation Action Plan was adopted and supported a pilot program in 28 cities.

The National New-Type Urbanization Plan (NUP, 2014–2020)[28] advanced more integrated green and inclusive urban development strategies and targets. This marked a paradigm shift from emphasizing economic growth to quality-oriented urban development. Urban areas now have specific targets: improved air and water quality, energy efficiency, solid waste management, water and wastewater management, and increased green spaces. These all aligned with the objectives of ecological civilization.

Partnering for a Sustainable Urban Future

The PRC–ADB partnership on urban development started in 1991. ADB's urban lending first focused on large cities and single sectors such as urban infrastructure and capacity development. Master plans for urban transport, water supply, wastewater management, and solid waste management focused on these areas. Later, central heating, river and lake rehabilitation, flood risk management, wetlands, and open space protection were added. ADB's urban technical assistance (TA) ranged from strategic policy advisory and technical guidance

28   The State Council of the People's Republic of China. 2014. National New Urbanization Plan (2014-2020).

on green and inclusive urbanization to urban environment improvement. This included management of water and wastewater, solid waste, construction and demolition waste, and flood risk. ADB's policy and advisory support helped design water and wastewater tariff reforms in the PRC and paved the way for public and private investments.[29] ADB's support ensured the financial and environmental sustainability of these operations.[30]

In recent years, the partnership in urban development shifted to more holistic approaches. Multisector projects started supporting the development of livable, green, and inclusive small and medium-sized cities more comprehensively, particularly in the less-developed regions of the PRC. ADB's partnership with the PRC from 2016 to 2020 is implementing several of the priorities of the NUP and 13th Five-Year Plan covering sponge cities, low-carbon cities, urban–rural integration, smart cities, and other innovative programs through loans and TA projects.

Several projects have aligned with ADB's Strategy 2030 and its operational priority on livable cities.[31] These projects now contain a higher level of integration across sectors— aiming at green and inclusive, low-carbon, and climate-resilient urban development. Projects have also been more directly supporting green transformation and green economic development. Technical education and training to build human resources for green jobs have been provided, geared toward retraining workers formerly engaged in dated carbon-intensive industries.

Demonstrating Proof of Concept

Several ADB projects have been supporting the revitalization and greening of urban areas in the PRC's northeast region. In 2014, ADB completed a large-scale program for environmental improvement and pollution control in the Songhua River Basin covering 32 cities in the two provinces of Heilongjiang and Jilin.[32] About 10 million urban residents in counties and cities benefited from the project, which provided wastewater treatment and clean and reliable water supply by developing new drinking water sources. The government adopted a basin-wide pollution control master plan prepared under a TA before the loan. The project was then simultaneously implemented with a private sector operations loan.

In Pingxiang municipality in Jiangxi province, ADB changed the conventional gray infrastructure approach of flood control, which channels the river with walls and hard embankments, to a sponge city green infrastructure approach.[33] More space and flow capacity is provided to the rivers. Floodplains and wetlands are maintained or rehabilitated, and ecological water edges allow for natural and seasonal fluctuations in water levels while also contributing to enhanced ecology and water quality. This is implemented in four townships— closer to rural residents—strengthening small cities and linkages with the core city.

About 2.73 million residents will now enjoy green and urban economic transformation

[29] ADB. 1999. *Completion Report: Water Tariff Study in the People's Republic of China.* Manila; ADB. 2002. *Completion Report: Water Tariff Study II in the People's Republic of China.* Manila; and ADB. 2003. *Completion Report: Preparing the National Guidelines for Urban Wastewater Tariffs and Management Study in the People's Republic of China.* Manila.

[30] R. Wihtol. 2018. *A Partnership Transformed: Three Decades of Cooperation between the Asian Development Bank and the People's Republic of China in Support of Reform and Opening Up.* Manila: ADB.

[31] ADB. 2018. *Strategy 2030—Achieving a Prosperous, Inclusive, Resilient, and Sustainable Asia and the Pacific.* Manila.

[32] ADB. 2015. *Completion Report: Songhua River Basin Water Pollution Control and Management Project in the People's Republic of China.* Manila.

[33] ADB. 2015. *Report and Recommendation of the President to the Board of Directors: Proposed Loan to the People's Republic of China for the Jiangxi Pingxiang Integrated Rural-Urban Infrastructure Development Project.* Manila.

through an ongoing ADB-financed project in the northeastern-most subregion of Heilongjiang province. Four coal and heavy industry-based cities will implement an environmental cleanup from mining impact and promote urban infrastructure. The project's support to green diversification will help promote the private sector.[34] Small and medium-sized enterprises (SMEs) engaged in non-coal sectors can enhance capacity and access to finance, leveraging further private sector capital. Two new projects are being prepared to continue support to the green transformation in Heilongjiang and Jilin provinces.

ADB promotes solid waste management in cities and river basins by reducing plastic pollution—contributing to ADB's new clean ocean initiative. An example of a project that includes an innovative solid waste management component is in Sichuan province, approved in 2018.[35] The project helped build flood control embankments, provided interventions to capture stormwater, and helped turn a polluted landfill into a park. A research and development center for light industries and a vocational training center were established to help local people acquire the necessary skills for future jobs. This transformative component of jobs reskilling helped the municipality improve its urban development planning and management.

In 2019, ADB approved a modern intelligent transport system (ITS) that will help reduce pollution, manage traffic flow, and improve transport safety in a fast-growing new city of Gui'an in Guizhou province.[36] The system comprises real-time monitoring and an integrated multimodal traffic operations, and safety, and emergency management system.

Through the project, local government staff will be trained in activities that will help make the infrastructure and services inclusive, gender-responsive, safe, and sustainable. A special zone will help share research and development knowledge on vehicle communication technologies in this key sector.

A comprehensive integration to transform a city to become greener and more livable is demonstrated in Jilin province (Box 5.1).

ADB has also been engaging the private sector in several urban projects especially in green activities including water and wastewater, and solid waste management (Appendix Table A3). In a project approved in 2020, ADB partnered with private sector companies to work with a number of third- and fourth-tier cities in the PRC to demonstrate the value of sponge city programs and natural infrastructure for water management (Box 5.2).

Guiding the Future

The following key lessons from urban planning and implementation of projects supporting the development of green and inclusive cities may guide future projects in this area.

First, systematic planning, integration, and effective coordination across sectors involving administrative departments, and local jurisdictions are required in implementing sustainable green urban livability. This requires integration of land-use planning, environmental management, and ecosystem protection, open space planning, infrastructure and services provision, sustainable mobility, and climate mitigation and adaptation, including disaster risk management.

---

[34] ADB. 2017. *Report and Recommendation of the President to the Board of Directors: Proposed Loan and Technical Assistance Grant to the People's Republic of China for the Heilongjiang Green Urban and Economic Revitalization Project*. Manila.

[35] ADB. 2018. *Report and Recommendation of the President to the Board of Directors: Proposed Loan to the People's Republic of China for the Sichuan Ziyang Inclusive Green Development Project*. Manila.

[36] ADB. 2019. *Report and Recommendation of the President to the Board of Directors: Proposed Loan to the People's Republic of China for the Guizhou Gui'an New District New Urbanization Smart Transport System Development Project*. Manila.

Box 5.1: Improving Urban Livability in Jilin Yanji

Low-carbon solutions for urban livability in Yanji. Sponge city green infrastructure and bus rapid transit corridor are integrated with green spaces for walking and cycling (photo by Guangzhou Municipal Design Institute).

In 2019, the Asian Development Bank (ADB) approved a $130 million equivalent loan to improve urban livability in Yanji through an integrated solution. The project focuses on public transport and promotes walking, cycling and healthy lifestyles and integrates sponge city green infrastructure as a platform for mitigation and adaptation. ADB finances the first bus rapid transit (BRT) corridor in the northeast of the People's Republic of China integrated with comprehensive stormwater management and water supply system improvements. The project design applies principles of transit-oriented development, focusing on higher density mixed-use and pedestrian-friendly center areas around BRT stations, promoting low-carbon urban mobility.

Around 590,000 people, including poor and low-income families, will soon enjoy new green spaces that are designed as sponge city green infrastructure link station areas, with project-supported riverfront greenways. The project pilots advanced computer modeling, demonstrating that integration of green and gray infrastructure can significantly reduce urban flooding (pluvial flooding).

Sources: ADB. 2019. *Report and Recommendation of the President to the Board of Directors: Proposed Loan to the People's Republic of China for the Jilin Yanji Low-Carbon Climate-Resilient Healthy City Project.* Manila; ADB. 2020. *Healthy and Age-Friendly Cities in the People's Republic of China: Proposal for Health Impact Assessment and Healthy and Age-Friendly City Action and Management Planning.* Manila.

Box 5.2: Increasing Climate Resilience Using Smart Urban Water Infrastructure

Smart urban water management in the People's Republic of China. At least 2 million more people, including women in urban areas, will benefit from new and improved urban water systems and facilities (photo by ADB).

To support the sponge city program, the Asian Development Bank (ADB) approved in 2020 a $200 million nonsovereign loan to expand climate-resilient and smart urban water management in the People's Republic of China (PRC). The loan, extended to the Shenzhen Water Group Co., Ltd. and Shenzhen Water and Environment Investment Group Co., Ltd., will help finance the construction and rehabilitation of urban water systems and facilities in selected cities in the PRC, where smart water knowledge and technology will be applied. At least 2 million more people, including women in urban areas, will benefit from urban water services.

Sponge city infrastructure has been shown to yield multiple benefits, including effective defense against flooding, better water absorption filtering and purifying, increased biodiversity, and a more conducive environment for recreation, walking, and cycling. ADB is supporting the transformation of traditional urban water utilities into comprehensive water environmental management providers and mobilize investment from the private sector to support sustainable development.

Source: ADB. 2020. *Report and Recommendation of the President to the Board of Directors: Proposed Loan to People's Republic of China for the Shenzhen Water (Group) Co., Ltd. and Shenzhen Water and Environment Investment Group Co., Ltd. Climate-Resilient and Smart Urban Water Infrastructure Project.* Manila.

Second, using nature and nature-based solutions creates synergies from a variety of ecosystems provided—benefiting both nature and people, human health, and economic competitiveness. On a local level, actions include diversified green open space system that connects with green city, neighborhood parks, and rain gardens with retention ponds to store stormwater.

Third, better and more sustainable urban mobility is another key component in designing green and more livable urban development. Urban transport must be integrated with land use and development planning and consider universal design principles to ensure inclusive and safe accessibility of public transport. Development partners in Asia and the Pacific can further support the development of regional ITS architecture with cross-border collaboration, including institutional and legislative agreements to help ensure investment returns and allow for knowledge sharing.

Lastly, increasing green and livable urban development also depends on increased partnerships with the private sector. The ability to act effectively and swiftly can be enhanced through institutionalizing coordination and engaging with the private sector. ADB continues to promote the rural–urban integration approach in various infrastructure sectors including public–private partnership (PPP) in sewage pipeline networks, centralized industry wastewater treatment, off-grid clean energy, liquefied natural gas, and further penetration of district heating.

Improving urban livability. ADB projects have been supporting the revitalization and greening of urban areas in the PRC (photo by ADB).

Access to clean energy. ADB's air quality improvement program in the Greater Beijing–Tianjin–Hebei region covers demonstration of clean energy-powered urban public transport in the PRC (photo by Chenxiao LI, China National Investment and Guaranty Corporation).

INVESTING IN CLIMATE CHANGE MITIGATION AND ADAPTATION

Addressing the Climate Change Challenge

The PRC has been true to its commitment in protecting global climate and becoming a serious advocate and a lead player in addressing climate change mitigation and adaptation. Under the Paris Agreement, the PRC aimed to peak CO_2 emissions by around 2030 and strives to achieve it earlier. By 2030, it expects to reduce CO_2 per unit of GDP by 60%–65% over the 2005 level, raise the share of nonfossil fuels in primary energy consumption to about 20%, and increase forest stock by around 4.5 billion cubic meters over 2005.[37] In September 2020, the government announced that the PRC will strengthen its 2030 climate target, its Nationally Determined Contributions, aim to peak its emissions before 2030, and achieve carbon neutrality before 2060.

The PRC is focused on promoting mechanisms to ensure the long-term sustainability of its efforts in lowering climate change impacts, such as establishing and developing the carbon market and pilot low-carbon cities,

low-carbon industrial parks and low-carbon community development, and supporting the research and development of cutting-edge technologies like the carbon capture, use, and storage (CCUS) technology. The PRC has taken the lead among developing nations to set-up a carbon market through the establishment of the national carbon emissions trading system. The government also developed and implemented a series of policies, programs, and actions that not only promote the domestic climate actions, but also laid a solid enabling policy environment for cooperation with international organizations like ADB.[38]

Partnering to Mitigate and Adapt to Climate Change Impacts

ADB has a long history of working with the PRC in addressing climate change. In 2015, ADB and the NDRC signed the MOU on Cooperation to Address Climate Change. This supports the PRC's efforts in achieving its climate change targets, particularly on GHG control in urban areas, low-carbon technology

---

[37] T. Fransen et al. 2015. A Closer Look at China's New Climate Plan (INDC). *World Resources Institute.* 2 July.

[38] The PRC's National Plan on Climate Change (2014-2020), National Strategy of Climate Change Adaptation, Energy Development during the 13th Five-Year Plan Period, Work Plan for Controlling Greenhouse Gas Emissions During the 13th Five-Year Plan Period, Circular of the State Council on the Issuance of the 13th Five-Year Plan for Eco-Environmental Protection, and Comprehensive Work Plan for Energy Conservation and Emissions Reduction During the 13th Five-Year Plan Period.

promotion, and carbon financing. On 27 May 2019, ADB and the Ministry of Ecology and Environment formalized and cemented this strategic partnership in promoting inclusive and environmentally sustainable growth in the PRC. ADB's contribution through this cooperation intends to generate knowledge through its projects, introduce international advanced experience, and promote knowledge sharing between the PRC and other ADB members in ecology, environment, and climate change.

ADB is also supporting the PRC to push forward low-carbon province and city pilots and to intensify the construction of low-carbon communities. For the agriculture and natural resources sector, ADB supports policy studies and reform that contribute to climate-smart agricultural practices and environmental management and protection. This strengthens the design and implementation of projects that focus on climate-smart agriculture, managing pollution for air, water, soil, and marine areas, including plastic waste, and conserving, protecting, and restoring natural habitats and ecosystem services (natural capital).

In 1992, ADB provided the first foreign technical support to the PRC in addressing climate change. This TA helped the government formulate the national response strategy for global climate change through policy advice and capacity building support. Through the assistance, the PRC established its climate change operational system, including policy making and advancing science and technology. Since then, multiple collaborations have been implemented, ranging from developing road maps for CCUS, providing

True to its commitment in protecting the global climate, the PRC will strengthen its 2030 climate target and its Nationally Determined Contributions, aim to peak its emissions before 2030, and achieve carbon neutrality before 2060.

inputs for the five-year plans, workshops and knowledge events, and policy-based loans that helped to improve air quality and mitigate GHG emissions.

ADB also provided technical and knowledge support in the development of the National Strategy for Climate Adaptation (NSCA) 2035, drafted by the Ministry of Ecology and Environment. ADB policy recommendations will be incorporated into the final document and form part of the national priorities on adaptation that will flow to the regional and local levels. ADB's policy advice will enhance and strengthen policy coordination, risk assessment for water-related disaster risks such as droughts and floods, climate finance, insurance, resilient infrastructure development, sustainable agriculture, management of rivers and oceans, and adoption of nature-based solutions such as river embankments and coastal protection as a buffer for flood control.

Demonstrating Proof of Concept

About 112 million individuals in the greater Beijing–Tianjin–Hebei (BTH) region have been worst affected by smog. ADB and the PRC established a lending program, spanning 2015 to 2020, for six loans amounting to $2.1 billion to reduce air pollution in the region (Figure 6.1 and Appendix Table A4). This assistance was later expanded to parts of Henan, Liaoning, Shandong, Shanxi provinces, and the Inner Mongolia Autonomous Region.

The first loan, approved in 2015, is a policy-based loan that strengthened regulatory and institutional frameworks on clean energy measures and investments in Hebei province and in the greater BTH area. It is the first ever completed lending assistance under ADB's support to the BTH region (Box 6.1).

The second and third loans (2016–2017) focused on the innovative use of financial

Figure 6.1: Asian Development Bank-Financed Loans on Air Quality Improvement in the Greater Beijing–Tianjin–Hebei Region, 2015–2020

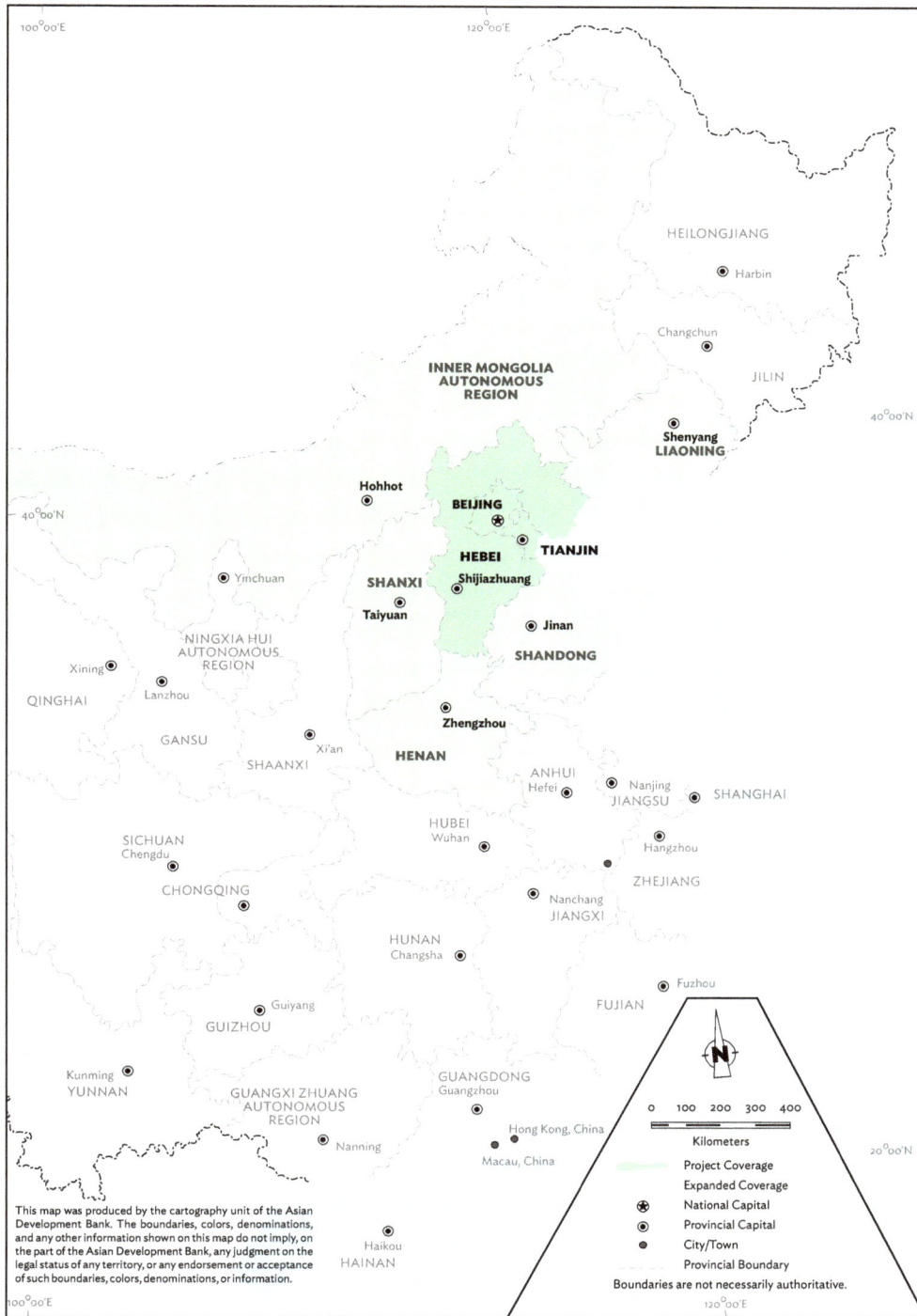

100°00'E   120°00'E

HEILONGJIANG

Harbin

Changchun

INNER MONGOLIA AUTONOMOUS REGION

JILIN

40°00'N

Shenyang
LIAONING

Hohhot

BEIJING

40°00'N

Yinchuan

HEBEI   TIANJIN

SHANXI   Shijiazhuang

Taiyuan

Jinan

NINGXIA HUI AUTONOMOUS REGION

SHANDONG

Xining   Lanzhou

QINGHAI

Zhengzhou

GANSU   Xi'an

SHAANXI   HENAN

ANHUI
Hefei   Nanjing
JIANGSU   SHANGHAI

SICHUAN
Chengdu

HUBEI
Wuhan

Hangzhou

CHONGQING   ZHEJIANG

Nanchang
JIANGXI

HUNAN
Changsha

Fuzhou

Guiyang

FUJIAN

GUIZHOU

N

Kunming

YUNNAN   GUANGXI ZHUANG AUTONOMOUS REGION

GUANGDONG
Guangzhou

0   100 200 300 400

Nanning

Hong Kong, China

Kilometers

Macau, China

20°00'N

Project Coverage

Expanded Coverage

National Capital

Provincial Capital

Haikou   City/Town

HAINAN   Provincial Boundary

Boundaries are not necessarily authoritative.

This map was produced by the cartography unit of the Asian Development Bank. The boundaries, colors, denominations, and any other information shown on this map do not imply, on the part of the Asian Development Bank, any judgment on the legal status of any territory, or any endorsement or acceptance of such boundaries, colors, denominations, or information.

100°00'E   120°00'E

Source: Asian Development Bank.

Box 6.1: Mitigating Air Pollution through Policy Reforms

Reducing air pollution in the greater Beijing–Tianjin–Hebei region. ADB is supporting new technologies to deliver clean heating and fuel supply (photo by Hebei Department of Finance).

The Beijing and Tianjin municipalities and the Hebei province comprise the Beijing–Tianjin–Hebei (BTH) region that is situated in the northern part of the People's Republic of China (PRC). Compared with the two municipalities, Hebei province was behind in terms of regional development. In 2015, about 88% of the 3.1 million poor people in the BTH region were from Hebei. As the country's second-largest coal consumer, the province accounted for more than 80% of primary particulate matter ($PM_{2.5}$) emissions in the BTH region. A weak environmental policy framework and incomplete set of regulations constrained the comprehensive control of air pollution in Hebei province.

To address this challenge, ADB approved a $300 million policy-based loan in 2015, which strengthened the framework for new environmental policies and investments in Hebei province. Through this program, the Hebei provincial government implemented 17 policy actions covering (i) reduction of air pollution in key industrial sectors, (ii) development of a comprehensive institutional framework for implementation of environmental policies and programs, and (iii) reemployment promotion for workers affected by the industrial transformation process.

Through the policy actions, Hebei reduced its annual coal consumption by about 12.4 million tons in 2017, representing about 4% of the province's total coal consumption in 2012. The provincial government committed to decommission 11,071 coal-based boilers and upgrade the remaining 23,562 units to meet higher energy-saving and environmental protection standards.

Efforts to further reduce production capacity in heavy-polluting coal-based industries are ongoing. In 2018, Hebei committed to further reduce its steel and iron production capacity and the unhealthy burning of biomass in the field. With the implementation of rigid environmental protection, monitoring, and penalties, seasonal biomass stalk burning was banned in Hebei. Three municipal-level cities and six county-level cities were selected as demonstration cities for clean-energy-powered urban public transport.

Overall, the policy reforms have helped cut air pollutants and greenhouse gas emissions substantially, with carbon dioxide emissions reduced by 18 million tons a year. From 108 micrograms/cubic meter in 2013, the annual average of $PM_{2.5}$ concentration in Hebei was reduced to 65 micrograms/cubic meter in 2017. Further reductions in annual pollutant emissions are expected.

The policy loan improved air quality regulatory enforcement capacity. In 2016, a law on air pollution prevention and control provided clear and binding provisions on volatile organic compound control, BTH air quality management coordination mechanisms, and accountability for environmental performance. The provincial government updated its air quality monitoring network and strengthened its air quality monitoring and evaluation system to effectively check and analyze ambient air quality. It also strengthened the foundation of resource development in Hebei by improving the quality of training institutions, trainers, and training programs for the successful redeployment and reemployment of workers.

Sources: ADB. 2018. *Completion Report: Beijing–Tianjin–Hebei Air Quality Improvement–Hebei Policy Reforms Program in the People's Republic of China*. Manila; Independent Evaluation Department. 2019. *Validation Report: Beijing–Tianjin–Hebei Air Quality Improvement–Hebei Policy Reforms Program in the People's Republic of China*. Manila: ADB.

intermediation to raise domestic funding for sustainable air quality improvement in the region. At the same time, capacities of local financial institutions were strengthened.

The second loan—the Green Financing Platform—targeted better access to finance, especially for SMEs, to scale up investments in pollution reduction projects.[39] SMEs account for about 60% of industrial pollution in the BTH region. The project provided credit guarantees to enable commercial financing from banks, debt financing through subloans, and financial leasing for purchasing energy-efficient industrial equipment for SMEs and energy service companies. The third loan established a regional emission-reduction and pollution-control financing facility for the greater BTH region.

In 2018–2019, ADB's multiyear program expanded to the highly polluted provinces in the region to support technology leapfrogging, and inclusive service delivery of clean heating and fuel supply. In Henan province, through a results-based lending modality, ADB will help support the switch from coal to gas, enhance institutional and organizational capacity to leverage sustainable funding for subsequent long-term investment, and finance numerous small-scale activities and expenditures across the province.[40]

The sixth loan, approved in 2020 for $150 million, will develop a credit enhancement scheme to support issuance of Clean Air Bonds aligned with international standard. This is the first air quality improvement dedicated green bonds in the PRC upon

issuance. It will also support cutting-edge energy technologies such as cooling system retrofit using water and ice—the first ADB intervention in the cooling sector in the PRC.[41]

The co-benefits of projects simultaneously reducing GHG, and other pollutant emissions have strong support from both the government and the public, which is critical for the success and sustainability of the projects. ADB's series of BTH loans demonstrates that investment projects can be designed to achieve co-benefits. ADB's overall support in this program revolves around strengthening policies and building the capacities of institutions, developing custom-fit financing models, and promoting advances in technology in key sectors and industries (that can reduce both air pollutants and GHG emissions).

As a result of ADB interventions and the Government of the PRC's actions, annual average $PM_{2.5}$ concentration has declined from 57 micrograms/cubic meter in 2015 to 40 micrograms/cubic meter in 2019, and annual average number of days in which air quality rated good to excellent has increased from 221 days in 2015 to 252 days in 2019, together with large GHG emission reductions as well as sulfur dioxide (SO_2), particulate matter, and nitrogen oxide (NO_x).

ADB is also demonstrating the use of green finance to expand clean technologies in SMEs (Box 6.2).

Using PPP to promote clean technologies in the PRC, ADB in 2009 approved the Municipal Waste-to-Energy Project to support the

[39] ADB. 2016. *Report and Recommendation of the President to the Board of Directors: Proposed Loan to the People's Republic of China for the Air Quality Improvement in the Greater Beijing–Tianjin–Hebei Region–China National Investment and Guaranty Corporation's Green Financing Platform Project.* Manila.

[40]   ADB. 2019. *Report and Recommendation of the President to the Board of Directors: Proposed Results-Based Loan to the People's Republic of China for the Air Quality Improvement in the Greater Beijing–Tianjin–Hebei Region—Henan Cleaner Fuel Switch Investment Program.* Manila.

[41] ADB. 2020. *Report and Recommendation of the President to the Board of Directors: Proposed Loan for the People's Republic of China for the Air Quality Improvement in the Greater Beijing–Tianjin–Hebei Region—Green Financing Scale up Project.* Manila.

Box 6.2: Leveraging Green Finance to Enhance Climate Resilience in Shandong Province

Financing infrastructure investments using green finance. ADB will help leverage private, institutional, and commercial finance for climate-resilient and low-emission investments (photo by ADB).

Approved by the Asian Development Bank (ADB) in 2019, the Shandong Green Development Fund Project enhances the climate resilience in Shandong province. By 2040, the project is expected to reduce carbon dioxide by 3.75 million tonnes, benefit 75 million people, and help to shift the investment in Shandong from "business-as-usual" to "transformational" climate-positive projects.

The project will establish a financing mechanism that will leverage private, institutional, and commercial finance for a pipeline of viable climate-resilient and low-emission investments.

This innovative mechanism is intended to be in a replicable and scalable form for future projects. The first innovative aspect is its financing mechanism. The $100 million ADB financing has mobilized $300 million from development partners to form a catalytic fund. Each development partner's financing will leverage private and institutional financing, leading to a funding size of $1.5 billion. This fund will translate into $7.5 billion investment, creating a fivefold multiplier effect. The second is linking climate impact with funding costs, with transformational project being provided with lower funding costs. The third is the integration between the project and knowledge.

The project was inspired by ADB's flagship publication, *Catalyzing Green Finance*. In turn, the project has generated several knowledge products to share its design features and experience.

The lessons from the Shandong Green Development Fund on green and leveraged finance approaches helped design the Association of Southeast Asian Nations (ASEAN) Catalytic Green Finance Facility (ACGF) in Southeast Asia that is helping green infrastructure projects transition across the funding gap so that private sector finance can help push green projects. The ACGF facility is helping ASEAN countries by providing infrastructure investments that are geared toward bankable and environmentally sustainable projects.

Sources: H. Jenny et al. 2020. Catalyzing Green Finance with the Shandong Development Fund. *ADB Briefs*. No. 144. Manila: ADB; ADB. 2020. *ASEAN Catalytic Green Finance Facility*. Manila; ADB. 2017. *Catalyzing Green Finance: A Concept for Leveraging Blended Finance for Green Development*. Manila.

construction and operation of waste–to–energy projects.[42] ADB structured a facility to support subprojects efficiently where the loan was provided to a holding company and channeled to the waste–to–energy project companies. The project has encouraged other local governments in Anhui, Fujian, Hunan, Qinghai, and Sichuan to pilot and implement PPP projects, replicating the business model. The knowledge and technology improvement learned from this project was also replicated in other countries.

ADB also supports low-carbon designs, climate-resilient infrastructure, developing smart city platforms, including intelligent transport to stimulate low-carbon behaviors and practices. In Xiangtan, ADB will demonstrate the transformative impact of inclusive transportation systems—complemented with information technologies, enabling policies, and institutional capacity building requirements for the city's low-carbon development.[43] In the PRC transport sector, ADB is guided by the

[42] ADB.2009. *Report and Recommendation of the President to the Board of Directors: Proposed Loan to the People's Republic of China for the Municipal Waste-to-Energy Project*. Manila.

[43] ADB. 2020. *Report and Recommendation of the President to the Board of Directors: Proposed Loan to People's Republic of China for the Xiangtan Low-Carbon Transformation Sector Development Program*. Manila.

"avoid–shift–improve" approach. This approach integrates land-use development with mobility needs to avoid the need for travel. It employs a shift to energy-efficient modes or routes of transport and seeks to improve vehicle and fuel technologies toward more energy-efficient ones.

Guiding the Future

ADB operations in the PRC have generated rich knowledge and experiences that have been and can be replicated in other provinces and cities as well as other DMCs to address climate change. Key experience and lessons learned to guide the future are summarized here.

First, strengthening the environmental policy framework and institutions can lead to comprehensive and inclusive efforts to address climate change mitigation and adaptation. ADB's first loan to help reduce air pollution in the BTH region improved Hebei province's enforcement and monitoring capacity. Following the success of the program, ADB provided further investments and technical support to improve air quality in BTH and its surrounding regions. A major lesson from the program is that sustained engagement by ADB is important in undertaking policy reforms that may take years to implement.

Second, capacity building also plays a pivotal role to support government in developing and implementing innovative new climate mitigation policies and instruments. ADB has actively collaborated with the PRC for climate policy and capacity building in advancing emerging technologies like CCUS.

Third, developing tailored financing approaches are key to unlocking and scaling up investments in pollution reduction projects. Along with supporting technology leapfrogging and inclusive clean energy service delivery, and mobilizing domestic financing across sectors—including energy, transport, urban development—agriculture plays a key role in reducing GHG emissions and in enhancing resilience. The use of fintech instruments for micro, small, and medium-sized enterprises could be explored, with the aim of providing tailored guarantee services for clean energy investments.

Fourth, investments in adaptation to climate change need to be scaled up to reduce the likelihood and impacts of water-related disaster risks in the PRC's major river basins (the Yangtze and the Yellow), which have been negatively affected by floods and droughts since the late 1990s. ADB policy work to support the national climate adaptation strategy 2035 will help forge the institutional and technical capacities for achieving the country's strategic goals on adaptation.

While the climate change canvas is vast, with complex issues that call for difficult decision making, with the right policies, sharing and learning from each other, taking a united global stance, and working in unison, the world can combat and control climate change. The PRC experience can make significant contributions to these global efforts leading up to the twenty-sixth session of the Conference of the Parties to the UN Framework Convention on Climate Change.

7

Healthy lifestyles. Locals keeping themselves fit in the park's exercise structures in Jiamusi, PRC (photo by ADB).

TAKING THE PARTNERSHIP FORWARD

Scaling Up Green Development

The PRC's 14th Five-Year Plan (2021–2025) heralds a time when environmentally sustainable green growth can outpace the old development patterns. It sets the stage for advancing green development which will help achieve important national and global goals for sustainable development established for 2030 and beyond. Such goals include the country's efforts under the Paris Climate Change Agreement and what might be needed for achieving a carbon neutral economy in less than 40 years.[44] It also includes the new Global Biodiversity Convention goals for 2030 being set at the COP 15 meeting in Kunming in 2021, which constitute the next critical step toward the CBD goal of a harmonious relationship between people and nature. The PRC's active role together with the international community will also be paramount for the climate talks leading to the COP 26 to be held in Glasgow in November 2021.

The PRC's objective of having the basic framework of ecological civilization in place by 2035 is a firm signal of its intent to continue a pathway of sustainable green growth,

with high-quality development as an outcome. The PRC–ADB partnership cases demonstrate how transformative change in thinking and action can take place in 5 to 10 years with noticeable results. The PRC and ADB have agreed to prioritize green development and focus on addressing climate change in their partnership over the coming 5 years, a direction already evident in many of the newer projects highlighted in the cases. ADB's new CPS for PRC, 2021–2025 and projects in the pipeline will contribute to scaling up green development in the country.

The aim for carbon neutrality has provided additional momentum to the environmental agenda. ADB operations in the PRC will focus on three interrelated strategic priorities: (i) environmentally sustainable development, (ii) climate change adaptation and mitigation, and (iii) an aging society and health security.[45] ADB's support will also include institutional strengthening, the generation of regional and global public goods, and the sharing of knowledge, especially with other DMCs.

Consistent with the CPS as well as the 14th Five-Year Plan, the PRC–ADB partnership on green development will continue to focus on the following areas.

---

[44] N. Stern and C. Xie 2021. *China's New Growth Story: Linking the 14th Five-Year Plan with the 2060 Carbon Neutrality Pledge.* London: Grantham Research Institute on Climate Change and the Environment, London School of Economics and Political Science.

[45] ADB. 2021. *Country Partnership Strategy: People's Republic of China, 2021–2025—Toward High-Quality, Green Development.* Manila.

Enhancing ecological protection and ecological conservation. ADB will continue to apply an ecosystem-based and climate-resilient approach to the YREB particularly in the Anhui Chao Lake basin. Based on experience gained from the YREB, ADB is designing a new program to promote high-quality green development in the Yellow River Ecological Corridor (YREC) (Figure 7.1).

Greening the rural economy. Forthcoming rural vitalization projects will promote green finance components to engage local commercial banks as intermediaries for lending to

eco-friendly SMEs and will introduce innovative approaches such as the use of block chain technology and green bonds. Setting up green funds to promote market-based mechanisms for eco-compensation will help protect the environment and generate additional livelihood opportunities in rural areas.

Developing green livable cities. Future urban development efforts will be anchored on a circular economy approach, including zero-waste cities. ADB will continue to support water reuse and efficiency improvement, balanced and low-carbon urban development,

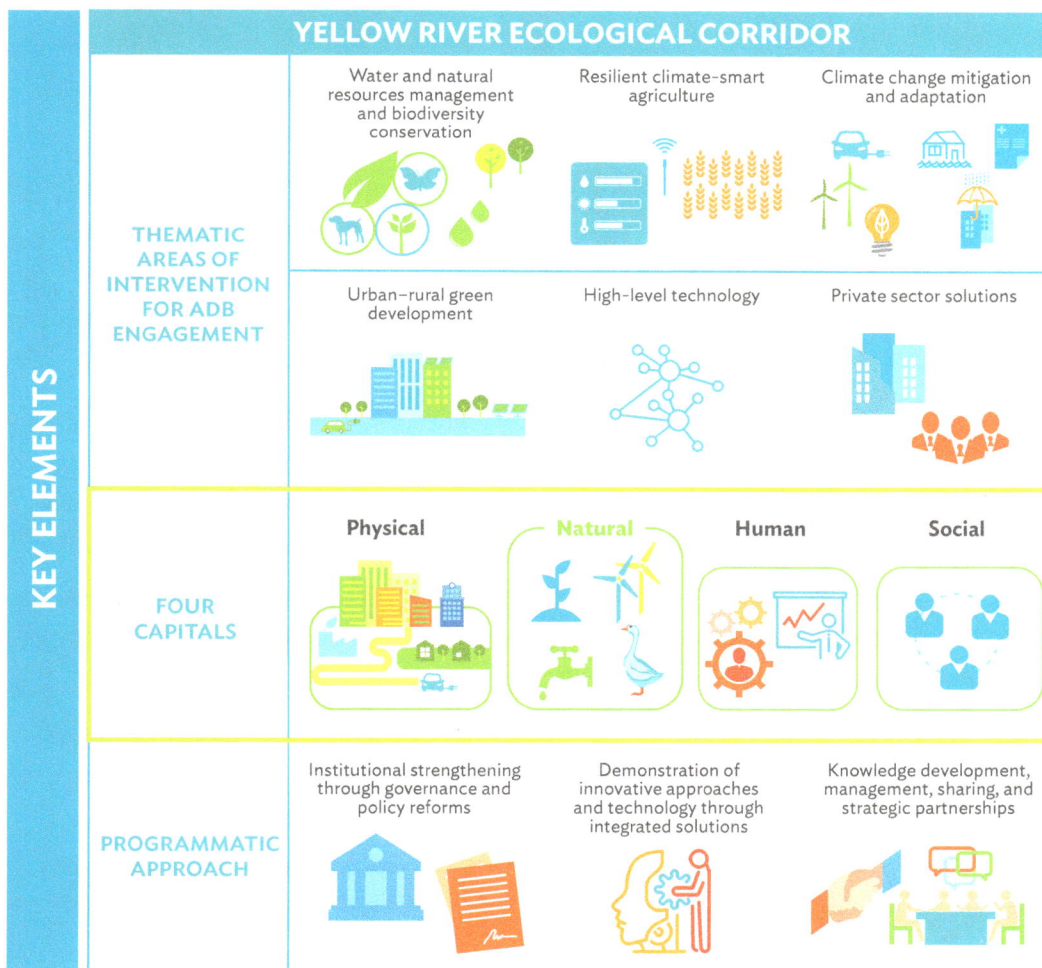

Figure 7.1: High-Quality Green Development in the Yellow River Ecological Corridor

ADB = Asian Development Bank.
Source: S. Cardascia and S. Robertson. Forthcoming. Scaling Natural Capital Investments in the Yellow River Ecological Corridor. Manila.

and nature-based solution for sustainable urban design as well as associated institutional and policy transformation, promotion of private sector finance, and advancement of technologies and innovations.

Investing in climate change adaptation and mitigation. ADB will remain to be a key partner with the PRC to scale up low-carbon development with more investments in innovative climate change solutions and strategic clean and low-carbon technologies such as CCUS, green hydrogen technologies, and digital technologies. ADB supports accelerated reforms to develop market instruments that treat carbon emission allowances and credits as assets.

Incorporating Green Finance to the Equation

Green finance needs to be an integral part of the solution to sustainable green development. New financial instruments and new policies, such as green bonds, green banks, carbon market instruments, fiscal policy, green central banking, financial technologies, and community-based green funds are collectively known as "green finance."[46]

In July 2020, the PRC launched the National Green Development Fund in Beijing. The fund will focus on key areas of green development along the YREB. More private capitals are expected to help in the country's ecological progress. The funds will be invested in fields such as pollution control, ecological restoration, afforestation of national land, conservation of energy and resources, green transportation, and clean energy.

As the PRC's role in the global green bond market expands, it is becoming increasingly important for the PRC to follow the internationally practiced taxonomy in green finance. In April 2021, the People's Bank of China released its updated taxonomy, the "Green Projects Catalogue," which includes a reference to the "Do No Significant Harm" principle. The catalog removed fossil fuel projects from the list, making it consistent with the European Union (EU) Taxonomy. The PRC–EU classification of green finance has inspired similar efforts around the world, from Colombia and South Africa to Singapore and Mongolia.[47]

The PRC has started to increase the role of regional banks to promote commercial banks to offer green finance products and services. The Bank of Xingtai project (Box 7.1) is such a pilot that is intended to be replicated and scaled up in other parts of the PRC.

Sharing the Knowledge

Knowledge is central to ADB's operations under the CPS and will provide a strategic approach for the PRC to share its development lessons with other ADB DMCs. The rich set of experiences with green development exemplified by the foregoing case studies, and many other projects, is readily accessible through different platforms. In addition, publications and other knowledge products are available on subjects such as green finance, eco-compensation, low-carbon efforts, and integrated water management, among others. Capturing and sharing this knowledge continues to increase understanding of green growth and development and highlight innovations in project design, technology, and management.

There remains a considerable need to fill knowledge gaps and to leverage

---

[46]   J. D. Sachs et al. 2019. *Why is Green Finance Important?* Tokyo: ADB Institute.

[47]   C. Jia. 2021. China, EU Lead Green Revolution with Finance Standards. *China Daily.* 29 April.

Box 7.1: Expanding Green Finance Lending with the Bank of Xingtai

Green finance with the Bank of Xingtai. ADB is supporting the establishment of a green finance lending business line, a replicable model of green finance development in the PRC (photo by Anqian Huang).

Approved by the Asian Development Bank (ADB) in 2020, the Bank of Xingtai Green Finance Development Project will help increase the volume of green finance provided by the city and rural commercial banking sector in Hebei province. The ADB funds will be provided to the Bank of Xingtai, a regional city commercial bank in Hebei, to finance subprojects that meet national and international green finance standards. Through this project, the Bank of Xingtai will, for the first time, adopt a green finance business line where each ADB dollar will mobilize at least $3 from the market.

The project will establish a green financing system for the Bank of Xingtai, including an advanced green finance information technology system that (i) certifies green subprojects; (ii) implements an environment, social, and governance (ESG) risk framework; and (iii) discloses ESG-related information. The project will introduce international best practices on green finance to the Bank of Xingtai and other city commercial banks, such as the European Union sustainable finance taxonomies, ESG, and Global Reporting Initiatives. It will also strengthen the Bank of Xingtai's institutional capacity in environment and social safeguards, gender mainstreaming, and financial management, and promote knowledge sharing in project implementation.

Sources: Government of the PRC, State Council. 2015. *The 13th Five-Year Plan for National Economic and Social Development, 2016–2020*. Beijing; ADB. 2020. *Report and Recommendation of the President to the Board of Directors: Proposed Loan to the People's Republic of China for the Bank of Xingtai Green Finance Development Project*. Manila.

partnerships in ways that can address specific issues and challenges on green development. The gaps include working more closely with businesses, which are the key source of innovations and can raise capital that ensure continuity in the spread of innovative solutions from small-scale to broad application. This is particularly the case for the various types of integrated, multisector projects. A unifying element is the need to draw upon information technology skills to a greater extent. These skills are key to marketing new ideas, building new relationships, and strengthening monitoring and knowledge-based platforms. Valuable partnerships already in place or under design include:

(i) China Council for International Cooperation on Environment and Development (CCICED).[48] CCICED Special Policy Studies have been carried out jointly by CCICED and ADB, with expert research teams defining and preparing policy advice. After review by the CCICED members, the key recommendations are circulated within the State Council and to relevant ministers and others. The most recent work (2018–2020) has been related to environmental protection topics in the YREB, including new innovations regarding eco-compensation and green development institutional reform.[49]

---

[48]   This high-level body provides advice to senior PRC leaders through the State Council. The chair is a PRC vice-premier. ADB vice-presidents have served as council members.

[49]   CCICED. 2019. *Special Policy Study on Ecological Compensation and Green Development Institutional Reform in the Yangtze River Economic Belt*.

(ii) Natural Capital Lab (NCL). Over a number of years there has been considerable collaboration between PRC national and international experts, for example a group within Stanford University and members in several PRC national academy groups. One outcome has been the design of ADB's NCL. The four main components of NCL include: (a) valuing natural capital; (b) strengthening policy, institutional, governance, and regulatory frameworks and tools; (c) catalyzing innovative financing mechanisms; and (d) building knowledge, capacities, and partnerships. The NCL will be launched as a digital platform in COP 15 in October 2021.

Preparing for the Future

The accumulated PRC experience from the recent past has already been shared internationally. More can be done, and in ways that will benefit both the PRC and other countries in the region. Some of the key concepts such as eco-civilization, eco-compensation, rural vitalization, redlining, and sponge cities are home-grown in the PRC. These are examples of how greening development has catalyzed new thinking and approaches that are also of interest to other DMCs.

The PRC–ADB partnership over the past few decades offer important insights for universal application in current and future initiatives. Moving quickly from theory to practical application and adaptive approaches is key to addressing emerging challenges in achieving environmentally sustainable and resilient development. At the core of the dialogue with the PRC will be the strengthening of environmental institutions and the supporting of policy reforms needed to advance the PRC's ambitious greening agenda.

ADB will continue aiming to create synergies in its green programs by designing projects bundling several green development objectives that mutually reinforce each other. While some initiatives bring immediate and apparent benefits to individuals, households, and small businesses, others may realize longer-term gains. Water supply and sewage or other waste management and public transport are good examples where such synergies would result in greater development impacts.

It is essential to consider digital platforms and other information technology applications in all projects. They knit the pieces together in bundled projects, and they are necessary in broadening communication and participation. Examples include complex technical matters such as reducing use of farm chemicals and water use, and for ecological conservation. An eco-regional approach helps build a good understanding of drivers, impacts, and innovative nature-based solutions. For example, the YREB efforts demonstrate value-added, and much promise ahead for integrated basin planning.

Green and sustainable finance is one key area in the coming years. ADB and the PRC will work together to reduce the significant gap between needs and available funding for sustainable projects. There is enormous potential for innovative green and blue finance to support structural transformation. Such shift will facilitate a sustainable low-carbon growth model, an enhanced circular economy (e.g., switching from brown to green energy, and expansion of carbon capturing), climate change resilience and adaptation (e.g., combating desertification), environmental sustainability (e.g., cleaner soil and water), and healthy oceans and sustainable use of marine resources (e.g., protect and restore coastal and marine ecosystems, rivers, and water resources). The green business models and technologies, and green financial institutions developed in the PRC can be replicated within and outside the Asia and Pacific region for a bigger development impact. For example, the PRC's role in co-chairing the sustainable finance working group of the G20[50] would facilitate such experience sharing and impactful replication.

To ensure funding continuity, scaling up from local sources will also be critical. The local green fund concept can reach out to rural and urban smaller communities with tested packages and funding for green initiatives. As such, packages become more attractive. They could address low carbon and zero emission problems in a more impactful manner. Also, innovations for conservation and ecological service enhancement might become more financially self-supporting through means such as regenerative agriculture and ecotourism.

Success in pursuing these initiatives rests largely on partnerships with various stakeholders to accelerate scaling-up and to seek more innovative approaches toward integrated, often spatially based, high-quality green development. Effective collaboration will take the partnership through different pathways: fully testing the potential of

new tools such as ecological redlining and eco-compensation; promoting the use of technological innovations; and bolstering further governance, policy reforms, and institutional and capacity strengthening. These pathways are needed to effect fundamental changes in the way environmental protection is managed in the PRC.

Green and sustainable rural development is the best way to secure ecological safety for all parts of the PRC. What the PRC does, or does not do, will have an immense impact on the future directions on climate change, ecological conservation, environmental and biodiversity degradation, balanced rural–urban development, and economic equality at the national, regional, and global levels.

ADB and the PRC will continue building on past joint ventures and new approaches to greening development for a more sustainable future. Piloting, dissemination, and replication of successful outcomes will be critical elements of the green development partnership between ADB and the PRC, bringing promising prospects to Asia and the Pacific, and beyond.

---

[50]   The G20 is an intergovernmental forum comprising 19 countries and the European Union. It works to address major issues related to the global economy, such as international financial stability, climate change mitigation, and sustainable development.

APPENDIX

Box A1: Eight Ecological Civilization Action Priorities in the People's Republic of China

1. Spatial planning and development

2. Technological innovation and structural adjustment

3. Land, water, and other natural resource sustainable uses

4. Ecological and environmental protection

5. Regulatory systems for ecological civilization

6. Monitoring and supervision

7. Public participation

8. Organization and implementation

Source: CCICED Chief Advisors. From Tipping Point to Turning Point. *CCICED Issues Paper 2012–2016*. 2014. Beijing: CCICED. pp 56-76.

Table A1: Asian Development Bank Sovereign Loan Approvals Supporting Green Development in the People's Republic of China, 2011–2020

Approval Year/Project Title	Amount[a] ($ million)
2011	**1,339.8**
Guangdong Energy Efficiency and Environment Improvement Investment Program – Tranche 3	42.9
Shandong Energy Efficiency and Emission Reduction	100.0
Forestry and Ecological Restoration in Three Northwest Provinces	100.0
Jiangsu Yangcheng Wetlands Protection Project	36.9
Qinghai Rural Water Resources Management Project	60.0
Guangxi Beibu Gulf Cities Development Project	200.0
Xinjiang Altay Urban Infrastructure and Environmental Improvement Project	200.0
Gansu Tianshui Urban Infrastructure Development Project	100.0
Xian Urban Road Network Improvement Project	150.0
Hai River Estuary Area Pollution Control and Ecosystem Rehabilitation Project	100.0
Railway Energy Efficiency and Safety Enhancement Investment Program – Tranche 3	250.0
Hebei Energy Efficiency Improvement and Emission Reduction Project	100.0

continued on next page

Table A1 *continued*

Approval Year/Project Title	Amount[a] ($ million)
2012	**1,470.0**
Hunan Xiangjiang Inland Waterway Transport Project	150.0
Comprehensive Agricultural Development Project	200.0
Jiangxi Fuzhou Urban Integrated Infrastructure Improvement Project	100.0
Heilongjiang Energy Efficient District Heating Project	150.0
Shanxi Energy Efficiency and Urban Environment Improvement Project	100.0
Hubei Huangshi Urban Pollution Control and Environment Management Project	100.0
Gansu Urban Infrastructure Development and Wetland Protection Project	100.0
Integrated Development of Key Townships in Central Liaoning	150.0
Ningxia Irrigated Agriculture and Water Conservation Demonstration Project	70.0
Anhui Chao Lake Environmental Rehabilitation Project	250.0
Shaanxi Weinan Luyang Integrated Saline and Alkaline Land Management Project	100.0
2013	**1,440.0**
Inner Mongolia Road Development Project	200.0
Railway Energy Efficiency and Safety Enhancement Investment Program – Tranche 4	180.0
Guangxi Baise Integrated Urban Environment Rehabilitation	80.0
Hubei Yichang Sustainable Urban Transport Project	150.0
Yunnan Sustainable Road Maintenance Project	80.0
Gansu Jiuquan Integrated Urban Environment Improvement Project	100.0
Xinjiang Integrated Urban Development	200.0
Chongqing Urban–Rural Infrastructure Development Demonstration II Project	150.0
Qinghai Delinha Concentrated Solar Energy Plant Project	150.0
Anhui Huainan Urban Water Systems Integrated Rehabilitation Project	150.0
2014	**1,440.0**
Railway Energy Efficiency and Safety Enhancement Investment Project – Tranche 5	170.0
Anhui Intermodal Sustainable Transport	200.0
Jiangxi Ji'an Sustainable Urban Transport Project	120.0
Yunnan Chuxiong Urban Environment Improvement Project	150.0
Yunnan Pu'er Regional Integrated Road Network Development Project	200.0
Jilin Urban Development Project	150.0
Hubei Huanggang Urban Environment Improvement Project	100.0
Gansu Baiyin Integrated Urban Development Project	100.0
Guangdong Chaonan Water Resources Development and Protection Demonstration Project	100.0
Low-Carbon District Heating Project in Hohhot in Inner Mongolia Autonomous Region	150.0

continued on next page

Table A1 *continued*

Approval Year/Project Title	Amount[a] ($ million)
2015	**1,579.0**
Shaanxi Mountain Road Safety Demonstration Project	200.0
Xinjiang Akesu Integrated Urban Development and Environment Improvement Project	150.0
Xinjiang Tacheng Border Cities and Counties Development	150.0
Henan Sustainable Livestock Farming and Product Safety Demonstration Project	69.0
Jiangxi Pingxiang Integrated Rural–Urban Infrastructure Development	150.0
Gansu Featured Agriculture and Financial Services System Development Project	100.0
Hubei Enshi Qing River Upstream Environment Rehabilitation Project	100.0
Chemical Industry Energy Efficiency and Emission Reduction Project	100.0
Hunan Dongjiang Lake Integrated Environmental Protection and Management Project	130.0
Qingdao Smart Low-Carbon District Energy Project	130.0
Beijing–Tianjin–Hebei Air Quality Improvement–Hebei Policy Reforms Program	300.0
2016	**1,599.6**
Shandong Groundwater Protection Project	150.0
Henan Hebi Qihe River Environmental Improvement and Ecological Conservation Project	150.0
Fujian Farmland Sustainable Utilization and Demonstration	100.0
Ningxia Liupanshan Poverty Reduction Rural Road Development Project	100.0
Chongqing Integrated Logistics Demonstration	150.0
Jiangxi Xinyu Kongmu River Watershed Flood Control and Environmental Improvement Project	150.0
Qinghai Haidong Urban–Rural Eco Development Project	150.0
Shaanxi Accelerated Energy Efficiency and Environment Improvement Financing Program	150.0
Air Quality Improvement in the Greater Beijing–Tianjin–Hebei Region—China National Investment and Guaranty Corporation's Green Financing Platform Project (formerly Green Financing Platform for Accelerated Air Quality Improvement in the Greater Beijing–Tianjin–Hebei Region)	499.6
2017	**1,629.0**
Mountain Railway Safety Enhancement Project	180.0
Shanxi Urban–Rural Water Source Protection and Environmental Demonstration Project	100.0
Shanxi Inclusive Agricultural Value Chain Development Project	90.0
Guizhou Rocky Desertification Area Water Management Project	150.0
Heilongjiang Green Urban and Economic Revitalization Project	310.0
Xinjiang Changji Integrated Urban–Rural Infrastructure Demonstration Project	150.0
Shandong Spring City Green Modern Trolley Bus Demonstration Project	150.0
Air Quality Improvement in the Greater Beijing–Tianjin–Hebei Region-Regional Emission Reduction and Pollution Control Facility	499.0

continued on next page

Table A1 *continued*

Approval Year/Project Title	Amount^a ($ million)
2018	**1,779.1**
Guangxi Regional Cooperation and Integration Promotion Investment Program - Tranche 2	180.0
Chongqing Longxi River Basin Integrated Flood and Environmental Risk Management Project	150.0
Yangtze River Green Ecological Corridor Comprehensive Agriculture Development	300.0
Yunnan Lincang Border Economic Cooperation Zone Development Project	250.0
Sichuan Ziyang Inclusive Green Development Project	200.0
Air Quality Improvement in the Greater Beijing–Tianjin–Hebei Region—Shandong Clean Heating and Cooling Project	399.1
Hubei Yichang Comprehensive Elderly Care Demonstration Project	150.0
Hunan Xiangjiang River Watershed Existing Solid Waste Comprehensive Treatment Project	150.0
2019	**1,530.2**
Multimodal Passenger Hub and Railway Maintenance Project	120.0
Demonstration of Guangxi Elderly Care and Health Care Integration and Public–Private Partnership	100.0
Gansu Internet-Plus Agriculture Development	130.8
Heilongjiang Green Urban and Economic Revitalization – Additional Financing	150.0
Shandong Green Development Fund	100.0
Air Quality Improvement in the Greater Beijing–Tianjin–Hebei Region--Henan Cleaner Fuel Switch Investment Program	300.0
Henan Dengzhou Integrated River Restoration and Ecological Protection Project	200.0
Anhui Huangshan Xin'an River Ecological Protection and Green Development Project	100.0
Jilin Yanji Low-Carbon Climate-Resilient Healthy City Project	130.0
Guizhou Gui'an New District New Urbanization Smart Transport System Development Project	199.5
2020	**1,494.8**
Chongqing Innovation and Human Capital Development Project	200.0
Yunnan Sayu River Basin Rural Water Pollution Management and Eco-Compensation Demonstration Project	100.0
Xiangtan Low-Carbon Transformation Sector Development Program	200.0
Jiangxi Shangrao Early Childhood Education Demonstration Program	100.0
Inner Mongolia Sustainable Cross-Border Development Investment Program (MFF Tranche 1)	196.3
Bank of Xingtai Green Finance Development Project	198.5
Hunan Miluo River Disaster Risk Management and Comprehensive Environment Improvement Project	150.0
Air Quality Improvement in the Greater Beijing–Tianjin–Hebei Region—Green Financing Scale-up Project	150.0
Shaanxi Green Intelligent Transport and Logistics Management Demonstration Project	200.0

^a Includes Asian Development Bank (ADB)-financed sovereign loans. Excludes investment grants and $100 million loan from Green Climate Fund which are fully administered by ADB.

Note: Projects that contribute to ADB's environmentally sustainable growth strategic agenda.

Sources: ADB loan, technical assistance, grant, and equity approvals database. ADB.

Table A2: Asian Development Bank Loans and Technical Assistance on the Yangtze River Economic Belt Program, 2015–2023

Project Title	Approval Year	Amount[a] ($ million)
Loans		
Jiangxi Xinyu Kongmu River Watershed Flood Control and Environmental Improvement	2016	150.0
Guizhou Rocky Desertification Area Water Management	2017	150.0
Chongqing Longxi River Environment Comprehensive Treatment and Ecological Protection Demonstration Project	2018	150.0
Sichuan Ziyang Inclusive Green Development Project	2018	200.0
Yangtze River Green Ecological Corridor Comprehensive Agriculture Development Project	2018	300.0
Henan Dengzhou Ecological Protection and Integrated Rehabilitation Project	2019	200.0
Hunan Xiangjiang River Watershed Existing Solid Waste Comprehensive Treatment	2019	150.0
Guizhou Gui'an New District New Urbanization Smart Transport System Development Project	2019	199.5
Yunnan Sayu River Basin Eco-Compensation Demonstration Project	2020	100.0
Anhui Huangshan Xin'an River Ecological Protection and Green Development Project	2020	100.0
Chongqing Innovation and Human Capital Development Project	2020	200.0
Hunan Miluo River Disaster Risk Management and Comprehensive Environment Improvement	2020	150.0
Hunan Xiangxi Rural Environmental Improvement and Green Development	2021	200.0
Jiangxi Ganzhou Rural Vitalization and Comprehensive Environment Improvement Project	2022	200.0
Anhui Chao Lake Comprehensive Environment Improvement, Phase II	2022	150.0
Chishui River Basin Ecological Protection and Green Development	2023	250.0
Technical Assistance		
Strengthening Provincial Planning and Implementation for the Yangtze River Economic Belt	2015	0.550
Guizhou High Efficiency Water Utilization Demonstration in Rocky Desertification Area	2016	0.500
Supporting the Application of the River Chief System for Ecological Protection in Yangtze River Economic Belt	2017	0.400
Preparing Yangtze River Economic Belt Projects	2017	1.500
Preparing Yangtze River Economic Belt Projects	2017	0.600
Preparing Yangtze River Economic Belt Projects (Supplementary)	2018	1.500
Preparing Yangtze River Economic Belt Projects	2018	1.583
Policy Research on Ecological Protection and Rural Vitalization for Supporting Green Development in the Yangtze River Economic Belt (four subprojects)	2018	1.500

continued on next page

Table A2 *continued*

Project Title	Approval Year	Amount[a] ($ million)
Piloting Innovative Flash Flood Early Warning Systems in Selected River Basins	2018	0.400
Policy Advice for the Yangtze River Protection Law of the People's Republic of China	2019	0.135
Capacity Building on River and Ocean Eco-Environment Management and Plastic Waste Pollution Control	2019	0.600
Developing Partnerships for Knowledge Sharing on Natural Capital Investment in the Yangtze River Economic Belt	2019	0.225
Innovating Eco-Compensation Mechanism in Yangtze River Basin (GEF)	2021	0.183
Innovating Eco-Compensation Mechanism in Yangtze River Basin (GEF)	2022	8.073

GEF = Global Environment Facility.

[a] Includes Asian Development Bank (ADB)-financed loans or ADB-administered grants for technical assistance.

Sources: ADB loan, technical assistance, grant, and equity approvals database. ADB. ADB East Asia Department.

Table A3: Asian Development Bank Nonsovereign Loan Approvals Supporting Green Development in the People's Republic of China, 2013–2020

Approval Year/Project Title	Amount[a] ($ million)
2013	**100.0**
Urban–Rural Integration Water Distribution Project	100.0
2014	**330.0**
Rural Smart Wastewater Treatment Project	100.0
Natural Gas for Land and River Transportation Project	150.0
Greenhouse Agricultural Development Project	80.0
2015	**250.0**
Western Counties Water and Wastewater Management Project	150.0
Small and Medium-Sized Enterprise Industrial Wastewater and Sludge Treatment Project	100.0
2016	**125.0**
Sustainable Dairy Farming and Milk Safety Project	62.5
Inclusive and Sustainable Livestock Farming Project	62.5
2017	**740.0**
China Everbright Renewable Energy Project	10.0
Green Transport Finance	200.0
Environmentally Sustainable Agriculture Input Distribution Project	80.0
Integrated Urban Water Management Project	200.0
Geothermal District Heating Project	250.0

continued on next page

Table A3 *continued*

Approval Year/Project Title	Amount[a] ($ million)
2018	**100.0**
Eco-Industrial Park Waste-to-Energy Project	100.0
2019	**100.0**
Industrial and Municipal Wastewater Treatment Project	60.0
Integrated and Sustainable Livestock Value Chain Project	40.0
2020	370.0
New Energy Bus Leasing Project	100.0
Solar Energy Finance Project	70.0
Climate-Resilient and Smart Urban Water Infrastructure Project	200.0

[a] Includes Asian Development Bank (ADB)-financed loans for nonsovereign projects that contribute to ADB's environmentally sustainable growth strategic agenda. The list includes projects that may have been terminated or cancelled post approval and excludes regional projects that may cover the People's Republic of China. Data available from 2013 only.

Sources: ADB loan, technical assistance, grant, and equity approvals database. ADB. ADB Private Sector Operations Department.

Table A4: Asian Development Bank Multisector Support Program for Beijing-Tianjin-Hebei Air Quality Improvement, 2015–2020

Project Title	Lending Modality	Approval Year	ADB Amount ($ million)	Financing for Climate Change Mitigation ($ million)	Estimated CO_2 Reduction (million tons per annum)
BTH Air Quality Improvement—Hebei Policy Reforms Program	Policy-based lending	2015	300.0	300.0	18.10
Air Quality Improvement in the Greater BTH Region—China National Investment and Guaranty Corporation's Green Financing Platform	Financial intermediation loan	2016	499.6	499.6	5.00
Air Quality Improvement in the Greater BTH Region—Regional Emission Reduction and Pollution Control Facility	Financial intermediation loan	2017	499.0	450.0	5.00
Air Quality Improvement in the Greater BTH Region—Shandong Clean Heating and Cooling Project	Stand-alone project	2018	399.9	360.0	3.99

continued on next page

Table A4 *continued*

Project Title	Lending Modality	Approval Year	ADB Amount ($ million)	Financing for Climate Change Mitigation ($ million)	Estimated CO$_2$ Reduction (million tons per annum)
Air Quality Improvement in the Greater BTH Region—Henan Cleaner Fuel Switch Investment Program	Results-based lending	2019	300.0	54.8	0.06
Air Quality Improvement in the Greater Beijing–Tianjin–Hebei Region—Green Financing Scale-up Project	Stand-alone project	2020	150.0	150.0	0.80

BTH = Beijing–Tianjin–Hebei, CO$_2$ = carbon dioxide.

Sources: Asian Development Bank (ADB) loan, technical assistance, grant, and equity approvals database. ADB. East Asia Department, ADB.

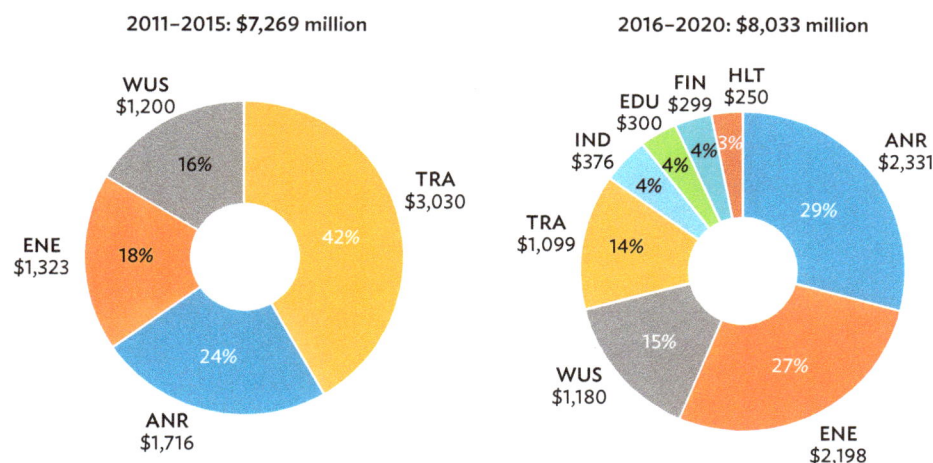

Figure A1: Asian Development Bank Approvals Supporting Green Development in the People's Republic of China by Sector, 2011–2020
($ million, %)

2011–2015: $7,269 million

WUS $1,200 — 16%
ENE $1,323 — 18%
ANR $1,716 — 24%
TRA $3,030 — 42%

2016–2020: $8,033 million

EDU $300
FIN $299
HLT $250
IND $376 — 4%
TRA $1,099 — 14%
WUS $1,180 — 15%
ANR $2,331 — 29%
ENE $2,198 — 27%
4%
4%
3%

ADB = Asian Development Bank, ANR = agriculture and natural resources, ENE = energy, FIN = finance, HLT = health, IND = industry and trade, TRA = transport, WUS = water and other urban infrastructure and services.

Note: Includes sovereign loans. Excludes $18.1 million investment grants financed by Global Environment Facility, and $100 million loan from Green Climate Fund, which are administered by ADB.

Source: ADB loan, technical assistance, grant, and equity approvals database.

REFERENCES

Asian Development Bank (ADB). 1999. *Completion Report: Water Tariff Study in the People's Republic of China*. Manila.

———. 2002. *Completion Report: Water Tariff Study II in the People's Republic of China*. Manila.

———. 2003. *Completion Report: Preparing the National Guidelines for Urban Wastewater Tariffs and Management Study in the People's Republic of China*. Manila.

———. 2009. *Report and Recommendation of the President to the Board of Directors: Proposed Loan to the People's Republic of China for the Municipal Waste-to-Energy Project*. Manila.

———. 2012. *Report and Recommendation of the President to the Board of Directors: Proposed Loan to the People's Republic of China for the Anhui Chao Lake Environmental Rehabilitation Project*. Manila.

———. 2012. *Report and Recommendation of the President to the Board of Directors: Proposed Loan to the People's Republic of China for the Comprehensive Agricultural Development Project*. Manila.

———. 2015. *Completion Report: Songhua River Basin Water Pollution Control and Management Project in the People's Republic of China*. Manila.

———. 2015. *Report and Recommendation of the President to the Board of Directors: Proposed Loan to the People's Republic of China for the Beijing–Tianjin–Hebei Region Air Quality Improvement Hebei Policy Reforms Program*. Manila.

———. 2015. *Report and Recommendation of the President to the Board of Directors: Proposed Loan to the People's Republic of China for the Jiangxi Pingxiang Integrated Rural-Urban Infrastructure Development Project*. Manila.

———. 2015. *Technical Assistance to the People's Republic of China for Modeling Urban Low-Carbon Development in Xiangtan*. Manila.

———. 2016. *Report and Recommendation of the President to the Board of Directors: Proposed Loan to the People's Republic of China for the Air Quality Improvement in the Greater Beijing–Tianjin–Hebei Region–China National Investment and Guaranty Corporation's Green Financing Platform Project*. Manila.

———. 2016. *Toward a National Eco-Compensation Regulation in the People's Republic of China*. Manila.

———. 2017. *Catalyzing Green Finance: A Concept for Leveraging Blended Finance for Green Development*. Manila.

———. 2017. *Report and Recommendation of the President to the Board of Directors: Proposed Loan to the People's Republic of China for the Air Quality improvement in the Greater Beijing–Tianjin–Hebei Region—Regional Emission-Reduction and Pollution Control Facility*. Manila.

————. 2017. *Report and Recommendation of the President to the Board of Directors: Proposed Loan to the People's Republic of China for the Guizhou Rocky Desertification Area Water Management Project*. Manila.

————. 2017. *Report and Recommendation of the President to the Board of Directors: Proposed Loan and Technical Assistance Grant to the People's Republic of China for the Heilongjiang Green Urban and Economic Revitalization Project*. Manila.

————. 2017. *Report and Recommendation of the President to the Board of Directors: Proposed Loan for the People's Republic of China for the Shandong Spring City Green Modern Trolley Bus Demonstration Project*. Manila.

————. 2017. *Technical Assistance to the People's Republic of China for Promoting Low-Carbon Development in Central Asia Regional Economic Cooperation Program Cities*. Manila.

————. 2018. *Completion Report: Beijing–Tianjin–Hebei Air Quality Improvement–Hebei Policy Reforms Program in the People's Republic of China*. Manila.

————. 2018. *Report and Recommendation of the President to the Board of Directors: Proposed Loan to the People's Republic of China for the Air Quality Improvement in the Greater Beijing–Tianjin–Hebei Region—Shandong Clean Heating and Cooling Project*. Manila.

————. 2018. *Report and Recommendation of the President to the Board of Directors: Proposed Loan to the People's Republic of China for the Chongqing Longxi River Basin Integrated Flood and Environmental Risk Management Project*. Manila.

————. 2018. *Report and Recommendation of the President to the Board of Directors: Proposed Loan to the People's Republic of China for the Sichuan Ziyang Inclusive Green Development Project*. Manila.

————. 2018. *Report and Recommendation of the President to the Board of Directors: Proposed Loan to the People's Republic of China for the Yangtze River Green Ecological Corridor Comprehensive Agriculture Development Project*. Manila.

————. 2018. *Strategy 2030—Achieving a Prosperous, Inclusive, Resilient, and Sustainable Asia and the Pacific*. Manila.

————. 2018. *Support for Rural Vitalization in the People's Republic of China*. Memorandum. 29 August.

————. 2018. *Technical Assistance to the People's Republic of China for Policy Research on Ecological Protection and Rural Vitalization for Supporting Green Development in the Yangtze River Economic Belt*. Manila.

————. 2019. *Report and Recommendation of the President to the Board of Directors: Proposed Results-Based Loan to the People's Republic of China for the Air Quality Improvement in the Greater Beijing-Tianjin-Hebei Region—Henan Cleaner Fuel Switch Investment Program*. Manila.

————. 2019. *Report and Recommendation of the President to the Board of Directors: Proposed Loan to the People's Republic of China for the Anhui Huangshan Xin'an River Ecological Protection and Green Development Project*. Manila.

————. 2019. *Report and Recommendation of the President to the Board of Directors: Proposed Loan to the People's Republic of China for the Gansu Internet-Plus Agriculture Development Project.* Manila.

————. 2019. *Report and Recommendation of the President to the Board of Directors: Proposed Loan to the People's Republic of China for the Jilin Yanji Low-Carbon Climate-Resilient Healthy City Project.* Manila.

————. 2019. *Technical Assistance Completion Report: Strengthening Provincial Planning and Implementation for the Yangtze River Economic Belt in the People's Republic of China.* Manila.

————. 2020. *ASEAN Catalytic Green Finance Facility.* Manila.

————. 2020. *Healthy and Age-Friendly Cities in the People's Republic of China: Proposal for Health Impact Assessment and Healthy and Age-Friendly City Action and Management Planning.* Manila.

————. 2020. *Report and Recommendation of the President to the Board of Directors: Proposed Loan to the People's Republic of China for the Air Quality Improvement in the Greater Beijing–Tianjin–Hebei Region—Green Financing Scale up Project.* Manila.

————. 2020. *Report and Recommendation of the President to the Board of Directors: Proposed Loan to the People's Republic of China for the Hunan Miluo River Disaster Risk Management and Comprehensive Environment Improvement Project.* Manila.

————. 2020. *Report and Recommendation of the President to the Board of Directors: Proposed Loan to the People's Republic of China for the Bank of Xingtai Green Finance Development Project.* Manila.

————. 2020. *Report and Recommendation of the President to the Board of Directors: Proposed Loan to the People's Republic of China for the Chongqing Innovation and Human Capital Development Project.* Manila.

————. 2020. *Report and Recommendation of the President to the Board of Directors: Proposed Loan to the People's Republic of China for the Hunan Xiangxi Rural Environmental Improvement and Green Development Project.* Manila.

————. 2020. *Report and Recommendation of the President to the Board of Directors: Proposed Loan to the People's Republic of China for the Shenzhen Water (Group) Co., Ltd. and Shenzhen Water and Environment Investment Group Co., Ltd. Climate-Resilient and Smart Urban Water Infrastructure Project.* Manila.

————. 2020. *Report and Recommendation of the President to the Board of Directors: Proposed Loan to the People's Republic of China for the Xiangtan Low-Carbon Transformation Sector Development Program.* Manila.

————. 2020. *Report and Recommendation of the President to the Board of Directors: Proposed Loan to the People's Republic of China for the Yunnan Sayu River Basin Rural Water Pollution Management and Eco-Compensation Demonstration Project.* Manila.

————. 2020. *Technical Assistance to the People's Republic of China for Strengthening Capacity, Institutions, and Policies for Enabling High-Quality Green Development in the Yellow River Ecological Corridor: Subproject 3: Green Farmland*. Manila.

————. 2021. *Completion Report: Comprehensive Agricultural Development Project*. Manila.

————. 2021. *Country Partnership Strategy: People's Republic of China, 2021–2025—Toward High-Quality, Green Development*. Manila.

————. 2021. *PRCM Observations and Suggestions: The 14th Five-Year Plan of the People's Republic of China—Fostering High-Quality Development*. Manila.

Cardascia, S. and S. Robertson. Forthcoming. Scaling Natural Capital Investments in the Yellow River Ecological Corridor. Manila.

China Council for International Cooperation on Environment and Development . 2019. *Special Policy Study on Ecological Compensation and Green Development Institutional Reform in the Yangtze River Economic Belt*.

Fitch Ratings. 2021. China Corporates Snapshot - December 2020: China's Green Bond Market to Stay Robust amid Policy Support. 23 December.

Government of the People's Republic of China, Ministry of Foreign Affairs. 2020. President Xi Jinping's Statement at the General Debate of the 75th Session of The United Nations General Assembly. 22 September.

Government of the People's Republic of China, The National People's Congress of the People's Republic of China. 2014. China Declares War Against Pollution. *Xinhua*. 6 March.

Government of the People's Republic of China, The State Council. 2014. *National New Urbanization Plan (2014–2020)*. Beijing.

Government of the People's Republic of China. 2020. *The Yangtze River Protection Law of the People's Republic of China*. 26 December.

————. 2015. *The 13th Five-Year Plan for National Economic and Social Development, 2016–2020*. Beijing.

Groff, S. 2018. Supporting PRC's "Mother River" Will Help Achieve Ecological Civilization. *Asian Development Blog*. 9 February.

Hanson, A. 2019. Ecological Civilization in the People's Republic of China: Values, Action, and Future Needs. *East Asia Working Paper No. 21*. ADB. Manila.

Hu, A. 2016. The Five-Year Plan: A New Tool for Energy Saving and Emissions Reduction in China. *Advances in Climate Change Research*. 7. pp. 222–228.

Huaxia. 2020. With Xiaokang Victory in Sight, China Saddles Up for Socialist Modernization. *Xinhua*. 31 December.

Huaxia. 2021. China's 14th Five-Year Plan Published in Booklet. *Xinhua*. 14 March.

Huaxia. 2021. Poverty Alleviation: China's Experience and Contribution. *Xinhua*. 6 April.

Independent Evaluation Department. *Validation Report: Beijing-Tianjin-Hebei Air Quality Improvement—Hebei Policy Reforms Program in the People's Republic of China*. Manila: ADB.

James, A. 2020. China Says It Has Met Its Deadline of Eliminating Poverty. *Wall Street Journal*. 23 November.

Jenny, H. et al. 2020. Catalyzing Green Finance with the Shandong Development Fund. *ADB Briefs* No. 144. Manila: ADB.

Jia, C. 2021. China, EU Lead Green Revolution With Finance Standards. *China Daily*. 29 April.

Jun, M. 2019. Five Insights Into China's Green Finance Transformation. *Central Banking*. 11 September.

Mu, X. 2017. China Outlines Roadmap for Rural Vitalization. *Xinhua*. 29 December.

Myllyvirta, L. 2019. Why China's CO_2 Emissions Grew 4% During First Half of 2019. *CarbonBrief*. 5 September.

Position Paper of the PRC for the UN Summit on Biodiversity. 2020. *Building a Shared Future for All Life on Earth: China in Action*. 21 September.

Sachs, J. D. et al. 2019. *Why is Green Finance Important?* Tokyo: ADB Institute.

Stern, N. and C. Xie. 2021. *China's New Growth Story: Linking the 14th Five-Year Plan with the 2060 Carbon Neutrality Pledge*. London: Grantham Research Institute on Climate Change and the Environment, London School of Economics and Political Science.

The People's Bank of China. 2016. *Guidelines for Establishing the Green Financial System*.

United Nations Environment Programme. 2021. *Making Peace with Nature: A Scientific Blueprint to Tackle the Climate, Biodiversity and Pollution Emergencies*. Geneva.

United States Geological Survey. Rivers of the World: World's Longest Rivers.

Wihtol, R. 2018. *A Partnership Transformed: Three Decades of Cooperation between the Asian Development Bank and the People Republic of China in Support of Reform and Opening Up*. Manila: ADB.

World Health Organization. Zoonoses.

WWF. 2020. *Living Yangtze Report 2020 Summary*.

Zhang, Q. and Crooks, R. 2012. *Toward an Environmentally Sustainable Future: Country Environmental Analysis of the People's Republic of China*. Manila: ADB.

Zhiming, N. 2020. *Development Asia Case Study: An Integrated Approach to Preserving the Wetlands of the People's Republic of China*. Manila: ADB. 12 August.

www.ingramcontent.com/pod-product-compliance
Lightning Source LLC
Chambersburg PA
CBHW042034220326
41599CB00045BA/7384